003$8.95

THE PESTICIDE CONSPIRACY

Books by Robert van den Bosch

THE PESTICIDE CONSPIRACY

BIOLOGICAL CONTROL (*with P. S. Messenger*)

SOURCE BOOK ON INTEGRATED PEST MANAGEMENT (*with Mary Louise Flint*)

Robert van den Bosch

The Pesticide Conspiracy

University of California Press
Berkeley · Los Angeles · Oxford

University of California Press
Berkeley and Los Angeles, California
University of California Press, Ltd.
Oxford, England

Library of Congress Cataloging-in-Publication Data

Van den Bosch, Robert.
The pesticide conspiracy / Robert van den Bosch.
p. cm.
Reprint, with new introd. Originally published:
Garden City, N.Y. : Doubleday, 1978.
Bibliography: p.
Includes index.
ISBN 0-520-06831-9. — ISBN 0-520-06823-8 (pbk.)
1. Pesticides industry—United States. 2. Pesticides—Environmental aspects—United States.
3. Agricultural chemicals industry—United States.
4. Agricultural chemicals—Environmental aspects—United States. I. Title.
HD9660.P33U59 1989
338.4′7668651′0973—dc20 89-4994
CIP

Royalties from the sale of this book go to the Robert van den Bosch Memorial Committee

Copyright © 1978 by Robert van den Bosch
Printed in the United States of America
All rights reserved
1 2 3 4 5 6 7 8 9

To my wife, Peggy, who cheerfully shares my tempestuous life—and to my mistress, the University of California, who makes that life possible.

FOREWORD

When this book first appeared in 1978, the jacket was decorated with an illustration of a cosmetically perfect red apple carved with a skull and crossbones. Ironically, in March 1989, as we were drafting this foreword for the reissue of the book, public controversy was storming over the use of the suspected carcinogen Alar® to enhance the appearance and storage qualities of apples and other produce. In his introduction to the book Professor van den Bosch chided the pesticide industry for its use of Madison Avenue gimmickry to sell pesticides to farmers. Eleven years later a *San Francisco Examiner* article (March 13, 1989) outlined a major chemical company's multi-million dollar advertising campaign to soft-sell pesticides to homeowners, using tranquil scenes of people enjoying bug-free gardens. The corporate targets and tactics to sell pesticides may be different, but the goal remains the same.

But what has happened in other areas covered by the book? It would be presumptuous to attempt to update Professor van den Bosch's book—it is unique. This foreword simply reviews some of the areas covered by the book from the perspective of recent events.

Nobel laureate Norman Borlaug retracted much of his assault on the concept of the balance of nature after realizing that modern plant breeding and heavy pesticide use are only a partial answer in modern pest control. Writing in 1988, Borlaug cautioned the new breed of biotechnologists who control center court that they may be overselling the scientific superiority of genetic engineering techniques for solving crop production and protection problems. Like Borlaug and others, the new technologists fail to assess their innovations in an agroecosystem context and ignore the documented ability of insects to meet and conquer all of humanity's prior direct challenges. Among the many new proposals is the incorporation of "bio-rational insecticides" into plants—an approach which, because of genetic adaptation in pests, will not provide the hoped-for definitive solution to pest problems.

What van den Bosch called the "pesticide treadmill" continues and expands into other areas of agriculture and in most regions of the world. In the third Robert van den Bosch Memorial Lecture Robert Metcalf pointed out that pesticide resistance in insects was increasing exponentially: by 1984, 447 species of arthropod pests worldwide were resistant to one or more pesticides. In addition, as predicted by van den Bosch, pesticide-induced outbreaks of target and nontarget pests are increasingly common as natural enemies are destroyed and the pests breed unchecked. A prime example of an induced pest problem is the rice brown planthopper in southeast Asia, which until recently was regarded as the major primary pest of rice. Because of this perception, it was the focus of a large program sponsored by international agencies to breed resistant plants. Only later did the scientific community learn, via Dr. Peter Kenmore, one of van den Bosch's students, that the pest was an insecticide-induced product of the "green revolution" itself. Very recently one of van den Bosch's academic grandsons, Dr. Kevin Gallagher, showed that genes for overcoming resistance to rice varieties not yet released were widespread in the pest population. Armed with some of this information, President Suharto of Indonesia banned the insecticides responsible for inducing outbreaks of this pest. This drastic government action resulted in no losses in yield. There are many other such examples.

There is still plenty of reason to be "eco-radical" in the sense van den Bosch meant it. Organisms (plants, dolphins, insects, etc.) are considered pests when they interfere with short-run economic interests, and there is a propensity to use quick fixes to solve these problems. On managed lands in developed countries, herbicide use in particular is soaring—with a concomitant widespread contamination of ground water. Forests in the northwest United States have been clear cut because it is thought more efficient and "economical"; besides, 2,4,5-T, albeit contaminated with dioxin, can suppress unwanted plant competitors during regeneration of the stand. The use of this chemical on Forest Service lands has been banned for public health reasons, but other chemicals have been substituted, and spraying continues on private lands. Managing the forest by selective logging is not considered a viable economic option, and forestry continues its headlong rush to become more like traditional agriculture—genetically narrow varieties of trees are planted and a heavy reliance on chemicals is developing to produce the crop with its attendant induced pest problems. Around the world the last of the primeval forests are being ravaged by corporations seeking cheap natural resources, by large and small farmers seeking new lands, and by corporate raiders seeking to pay off huge debts. Unfortunately, making a fast buck still overrides all other considerations.

DDT use is no longer a major issue in developed countries, but its legacy lives on in the biosphere as new pesticides and other compounds of all kinds are introduced by industries at an alarming rate. Only later will we know some of the health or biological consequences of a small fraction of these compounds. Who is there to protect the public's interest? Certainly not the Environmental Protection Agency! We only need to consider what the Reagan years in Washington did to an already "raped" EPA and to the environment. Love Canals and Kesterson Reservoirs are more common than we would like to think. And consider the lovely Rhine, purged of life by a massive accidental Sandoz spill—compounded, we are told, by the purposeful subsequent dumping by others of chemicals they needed to dispose of. After all, the river had already been contaminated! More fearsome is the unknown fate of the myriad of chemical products and by-products dumped legally in landfills that have proven unsafe, and illegally in innumerable places far from sight. Unabashed, humans continue to pollute with gross lack of concern for the welfare of Mother Earth or of our future generations.

And what is new with "Ole King Cotton," that major abuser of pesticides? In southern California cotton growers sought and received a special use permit for a banned suspected carcinogenic pesticide—not because it was an effective insecticide but rather because farmers thought it enhanced yields by stimulating plant growth. Van den Bosch predicted that excessive pesticide use would simply exacerbate pest problems in cotton. In this regard, insecticide resistance in induced pests brought financial ruin to growers in the desert valleys of southern California; induced pests disgraced a major international pesticide company in the Sudan; and insect pressure forced Arkansas farmers to cooperate, to treat pests as a regional problem, and to substitute sound information on natural controls for indiscriminate use of insecticides by individual farmers—without losses in yield or quality. Now the dreaded cotton boll weevil has been introduced into Brazil, where it is devastating the cotton crop and reshaping the agricultural economy just as it once did in the southern United States. Brazilian plant breeders at first were confident, despite all historical evidence to the contrary, that they could control the boll weevil with insecticides. But a socially conscious Brazilian colleague knew differently and lamented, "They have poisoned the grandfathers and the fathers, and with the introduction of this pest they will also poison the sons." Most of the pesticides in developing countries continue to be applied using hand-held sprayers without the benefit of protective clothing, and in the United States this danger has only partially abated.

Professor van den Bosch thought that a better way to "battle bugs" was via the implementation of sound agroecological research (Integrated Pest Management) that substitutes the abundant natural controls of pests that exist in many agriculture systems for chemical controls. The seeds he and colleagues planted are now bearing fruit as prior knowledge of pest biologies is compiled, new knowledge is added to the literature, and computer-based systems are developed to deliver sound pest-control information. Although the increasing public awareness of the adverse side effects of indiscriminate pesticide use was of immense help in promoting the development of IPM and other research programs, the fact that pesticides are "too cheap" greatly hinders further development and implementation of IPM. The real costs of pesticides and other chemicals are not borne by the user: profits from pesticide use are private, but the negative health and environmental costs of their use remain in the public domain—in the water we drink, the food we eat, and the air we breathe.

Van den Bosch assailed CAST (the Council for Agricultural Science and Technology) and the giveaway "free pesticide press" for telling the "truth" about pesticides. Unfortunately, CAST is still alive and well; one of our graduate students recently received one of their mailers. The free pesticide press continues to be an important source of "information" for farmers. A Fall 1987 article in *The Cotton Grower* called Integrated Pest Management in cotton a nine-billion-dollar mistake, and the "instant professionals" loved it because they could now justify the use of more insecticides to save every last cotton bud—this even though the scientific community knows that cotton under pest-free conditions can mature less than half of the fruit initiated. Van den Bosch would have enjoyed the battle to set this and other scores straight, but alas he is gone.

How are the small farmers doing—van den Bosch's "sorriest losers"—who receive all the free information concerning the supposed benefits of pesticide use? Well, they will be unhappy to learn from the rural sociologists that the San Joaquin Valley continues to develop into the bastion of corporate farming, that their stable farm communities are being replaced by Spanish-speaking ones, that the welfare rolls for the corporate agricultural rich and the disenfranchised poor are on the rise, that the 160-acre rule for the use of public irrigation water has been changed to 960 acres but nobody enforces the new limit either, and that the role of the land-grant universities in promoting rural life has all but disappeared in California.

And what has happened to the farm-worker leader Cesar Chavez, whom van den Bosch supported so strongly? He nearly died last year during a prolonged fast designed to focus attention on the continuing dire plight of farm workers. Similarly, in Fall 1988 the co-founder of the United Farm Workers Union, Dolores Huerta, had four ribs broken and her spleen ruptured (on national television) by police using "prescribed" methods of crowd control. "Sticking it to Cesar" was van den Bosch's euphemism for keeping the downtrodden farm worker down, and nothing has changed in that area.

The politics of pest control also continue as usual. Even van den Bosch's professed "mistress," the University of California, proposes that corporate–university relationships for the development of biotechnology should be strengthened. Many applications of biotechnology will have important consequences for agriculture and pest control. The fear among many concerned researchers is that proprietary rights will supersede the public's right to know, that more and more the means of production will be vertically integrated, that unforeseen technology-driven pest-control disasters loom ahead, and that the genetic diversity of crops will be further reduced and controlled in the name of corporate profit and extra monies to run the university. These dire prospects augment the insult to the academic integrity of a great public land-grant university and to the academic freedom of its researchers.

Professor van den Bosch was one of the world's foremost experts in the field of biological control, and he no doubt would be saddened by recent events that affect his beloved discipline. The institutionalization of biological control is such that the term has been cheapened—it has lost its meaning. Everyone claims to be doing it, but the discipline is weaker now than it has ever been. This problem is best illustrated by a 1988 report from the prestigious National Academy of Sciences, wherein biological control was equated with biotechnology. In rebuttal, R. Garcia et al., in a 1988 *Bioscience* article, pointed out that this was a gross perversion of an honored concept and discipline. To make matters worse, the Division of Biological Control at the University of California at Riverside was disbanded in 1988, and the fate of the small group at Berkeley is uncertain. All of this occurs as we celebrate the centennial of the beginning of modern biological control: the introduction of the Vedalia beetle in 1888–89, which saved the citrus industry in California from the ravages of the cottony cushion scale. Yet as this is happening, widespread public outrage is demanding safer methods of pest control, and of course Professor van den Bosch's legacy outlined in this book urges us on. Such travail would merely have reinforced van den Bosch's resolve to preserve and strengthen the discipline of biological control, to maintain its identity. We must not fail in this task!

His wife Peggy and we, his colleagues in life, applaud

the University of California Press for reissuing this historical document that recounts one man's fight to save Mother Earth. And a bitter fight it was! "Van" fought with a conviction supported by his scientific endeavors, with the energy motivated by the pervasiveness of the problem, and with the passion inspired by his profound love for his academic institution, the University of California. The book bears dated facts, but readers will find its strong message as relevant today as when it was first issued.

L. E. Caltagirone
D. L. Dahlsten
L. K. Etzel
L. A. Falcon
R. Garcia
A. P. Gutierrez
K. S. Hagen
C. B. Huffaker,
Emeritus Professor
G. O. Poinar Jr.
Y. Tanada,
Emeritus Professor

Division of Biological
Control
University of California,
Berkeley

E. F. Legner
R. F. Luck
J. A. McMurtry
V. M. Stern
Paul De Bach,
Emeritus Professor

Formerly Division of
Biological Control
University of California,
Riverside

L. A. Andres

Formerly USDA/Biological Control of Weeds Laboratory,
Albany

PREFACE

With the publication, in 1962, of *Silent Spring*, highlighting the potential for ecological disaster inherent in the wide use of pesticides, Rachel Carson started the world down the road to ecological awareness. But in certain circles, that valiant writer's efforts have remained anathema. Pesticides were big business in 1962 and are still big business, and pesticides are an ideal product: like heroin, they promise paradise and deliver addiction. And dope and pesticide peddlers both have only one cure for addiction: use more and more of the product at whatever cost in dollars and human suffering (and in the case of pesticides, in environmental degradation).

The big-money moguls in the pesticide industry, their wholly-owned subsidiary the U. S. Department of Agriculture, their bought-and-paid-for entomologists and toxicologists, and the poor slobs who try to make a living promoting broadcast use of pesticides didn't like *Silent Spring* one bit. More than a decade after Dr. Carson's death, they still on occasion revile her. They will like Robert van den Bosch's book even less, for the book tells the public for the first time what competent professionals in the insect-control business have long known: that even without considering the environmental hazards of pesticides, *their broadcast use is a disaster for all but those who sell or promote them*. The pesticide system of today doesn't control pests, it creates them. It imposes an immense financial burden on farmers and an immense health burden on farm workers. And it exposes consumers to unknown risks with no compensatory benefits.

Professor van den Bosch, a distinguished scientist and "insider" with long experience in the business of controlling pest populations, lays out the story of stupidity, venality, and corruption as only an insider can. It's all here: the suppression of research on alternative systems, the sale of the honorable traditions of the Entomological Society of America for a mess of booze, the pressure put on scientists in state universities to suppress results unfavorable to the "pesticide mafia," the disgrace of the Department of Agriculture, the rape of the EPA—the whole tragic story.

The Pesticide Conspiracy is a book written by a man who is frankly angry, and you will be angry when you've finished it. But anger is not enough. Concerted political action is required if the desperately needed transition to integrated pest management is to be achieved. With such management, pesticides are used *when needed* in an ecologically sound mix of techniques that minimize damage to the crop, the environment, the farm worker, the consumer, and the farmer's pocketbook. Until the Agricultural Research Service of the Department of Agriculture can be upgraded and the USDA as a whole brought into the battle against pests (rather than in favor of pesticides), there is no hope of reform. And the quickest way to get to the USDA is by weeding out its overlords on the House Agriculture Committee—many of whom should have been retired decades ago for the benefit of the nation. As long as such men—ignorant of ecology but having enormous power over agriculture—remain in office, the agro-ecosystems of the United States will continue to run downhill toward ultimate disaster.

Paul R. Ehrlich
Professor of Biological Sciences
Stanford University

PREFACE TO THE 1989 EDITION

I'm sad to say that, having reread the preface I wrote in 1978, I see no reason to change its overall thrust. However, my subsequent experience with the United States Department of Agriculture, in connection with the Medfly disaster, does lead me to believe that I perhaps held too high an opinion of it in 1978.

Paul Ehrlich

PROLOGUE

Silent Spring Revisited

On this day in a future time, spring has returned and the sun-washed Mississippi Delta is breathtakingly beautiful under a brilliant blue sky and a mantle of lush greenery. It is a glorious season of warm days and soft nights, when the land, as it has for countless centuries, pulses with renewed life. There are the sounds of spring too: the rustling of leaves in a zephyr, the faint rumble of distant thunder, the whisper of water in a slowly flowing stream, the chitchat of small birds, the chirring of cicadas, the crackling flight of grasshoppers, the low hum of foraging bees.

But these are the voices of Nature, and they form a muted chorus compared to the crescendo of sound once heard in this vibrant season. Something is missing from this springtime symphony; that something is the sound of man. Man is gone, and nowhere to be heard is the clatter of a tractor, the moaning power surge of a spray plane, the grinding throb of a diesel locomotive, the roar of a truck-trailer rig, the squawk of a transistor radio, the whooping of voices across a field.

At some time in the past, man fashioned a catastrophe

and vanished in it. Now all that is left of his brilliant doings are the decaying remnants of the things he created. Ghosted cities and towns crumbling to rubble, automobiles rusted to junk on littered streets, farmhouses and outbuildings rotting to dust, croplands tangled with brush and invaded by forest, graveyards with headstones toppled among the weeds.

Yes, man has vanished, but the insects still abound, and on this lovely day theirs is the prevailing voice of an almost silent spring.

A fable?

Let's talk about it.

THE BUG BOMB

Chapter 1

INTRODUCTION

A Can of Worms

In the early summer of 1976, a popular California radio station broadcast to growers an insecticide advertisement prepared for a major chemical company by a New York ad agency. The broadcast warned the growers of the imminent appearance of a "menacing" pest in one of their major crops and advised that as soon as the bugs "first appear" in the fields the growers should start a regular spray program, using, of course, the advertised insecticide. The broadcast also claimed that the material was *the one* insecticide the growers in the area could depend on for effective and economical control of the threatening pest, and further told the growers that through its use they would get a cleaner crop and more profit at harvest time.

The advertisement epitomizes what is wrong with the American way of killing bugs, a practice more often concerned with merchandising gimmickry than it is with applied science. In connection with this gimmickry, much of modern chemical pest control is dishonest, irresponsible, and dangerous. This was true of the radio advertisement just described. It was *dishonest* in its claim that the touted insecticide was *the one* material that growers could depend upon, for in actuality there are several equally effective insecticides and none will assure a cleaner crop and more profit. The advertisement was *irresponsible* in advising growers to initiate a regular spray program upon "first appearance" of the pest. Intensive research has shown that spraying of the crop should be undertaken only when the pest population reaches and maintains a prescribed level during the budding season and that sprays should never be applied on a regular schedule. Finally, the advertisement was *dangerous,* because the advised spraying, if widely adopted by the growers, would have resulted in the senseless dumping of huge amounts of a highly hazardous poison into the environment.

As a veteran researcher in insect control, I have long been disturbed by the dishonest, irresponsible, and dangerous nature of our prevailing chemical control strategy, but I am even more distressed by the knowledge that this simplistic strategy cannot possibly contain the versatile, prolific, and adaptable insects. For a third of a century following the emergence of DDT, we have been locked onto this costly and hazardous insect control strategy, which for biological and ecological reasons, never had a chance to succeed.

What is most disturbing of all is our inability to clean up the mess by shifting to the workable, ecologically based, alternative strategy that modern pest-control specialists term *integrated control* (also termed integrated pest management). Integrated control, as the name implies, is a holistic strategy that utilizes technical information, continuous pest-population monitoring, resource (crop) assessment, control-action criteria, materials, and methods, in concert with natural mortality factors, to manage pest populations in a safe, economical, and effective way. Integrated control is the only strategy that will work effectively against the insects, because it systematically utilizes all possible tactics in such a way that they attain full individual impact, function collectively for maximum mutual effect, and cause minimum detriment to the surrounding environment. In other words, unlike the prevailing chemical control strategy, with its emphasis on product merchandising, integrated control is a technology. It is scientific pest control and, as such, the only way we can hope to gain the upper hand in our battle with the insects. In every respect, integrated control makes sense, and it works (see Chapter 15). Despite this, our swing to this better pest-management strategy has been painfully slow, and for a clear reason. The impediment has been a powerful coalition of individuals, corporations, and agencies that profit from the prevailing chemical control strategy and brook no interference with the status quo. This power consortium has been unrelenting in its efforts to keep things as they are and as so frequently happens in our society, the games it plays to maintain the status quo are often corruptive, coercive, and sinister.

This book, then, is a tale of a contemporary technology gone sour under the pressures generated by a powerful vested interest. Bugs provide the theme, but politics, deceit, corruption, and treachery are its substance. I feel that the story is a most timely one, for it describes an ecological ripoff and how this atrocity is being perpetuated by tacticians of pure Watergate stripe. The book is largely based on personal experiences and insights gained from more than a quarter century of battling the bugs and their human allies who devised and maintain the inadequate chemical control strategy. It is a tale of personal outrage that I hope proves highly infectious.

Chapter 2

MOTHER NATURE AND THE GRAY COMPUTER

The overwhelming tragedy of planet Earth is man's contempt for nature. And nowhere is this contempt more manifest than among agri-technologists, the architects of modern crop production. Listen to Norman Borlaug, father of the Green Revolution, as he voices this contempt: "The cliché 'in balance with nature' which is in common usage today by modern-day environmentalists is very misleading. It implies we would have a favorable 'balance with nature' to assure the protection of our crop species if the 'balance with nature' were not upset by man. This, of course, is not true, nor is there in existence a single in-balance-with-nature ecosystem."

This astonishing statement was made by Nobel Laureate Borlaug in his 1971 McDougall Memorial Lecture before the Seventh Biennial Conference of the Food and Agriculture Organization of the United Nations, in Rome.[1] Clearly, Borlaug fails to comprehend that the Green Revolution, his vehicle to international fame, became an imperative because one of earth's creatures had fallen out of balance with nature. This creature, of course, is man. It is cause for deep concern that people like Borlaug have immense influence over the way humanity meets the crisis of its own doing, and if these experts do not even understand nature's basic mechanisms, their advice and activities can only aggravate the earthly tragedy.

This is certainly true of my branch of technology, insect control, in which Borlaug stands steadfastly with those who support the chemical control strategy, an approach that has triggered a veritable Bug Bomb. Indeed, that is what this book is all about: technology's ironic role in enhancing the insects' competitive position vis-à-vis man. It is difficult to envisage how such lowly creatures could attain prime status as our competitors for the earthly bounty, but somehow our peculiar genius for disrupting nature has boosted them into that position.

Let me tell you about it.

The Insects

As a group, insects are the most successful animals that have ever evolved. For one thing, they are ancient, having appeared more than 300 million years ago. Furthermore, by 200 million years ago they had diversified remarkably and had developed forms very similar to those running and flitting about today.[2] What this tells us is that insects have held their own over a vast expanse of time in the face of cataclysmic geological, climatic, and biological changes which have wiped out untold numbers of more "advanced" creatures.

In other words, insects have a fantastic ability to survive. Why? Well, for one thing, they are incredibly diverse. Various estimates place their total species' numbers in excess of 1 million. Roughly 75 per cent of all of Earth's described animal species are insects.[3] Insects are also highly adaptable; they are literally everywhere, occupying an amazing array of niches, embracing water and soil, the roots, stems, branches, leaves, flowers, and fruits of plants, stored plant and animal products, furniture, clothing, books, house foundations, fence posts, dung, carrion, and living animals, including one another, dogs, cats, rats, squirrels, wolves, sea lions, and humans, to name but a few.

Many insects also have incredible reproductive abilities, which are realized through fantastic egg-laying capacities or by the rapid turnover of generations. The champion egg layer is probably the termite, which pumps out about 150 million eggs during her fecund life span. The quick-cycle artists are such creatures as the aphids, which under optimum conditions can turn over a generation per week. The numbers that the insectan birth machines grind out are truly mind-boggling. For example, a locust swarm may cover six hundred square miles and contain more than a trillion insects.[4] Each year, in California alfalfa fields alone, 7.5 billion convergent lady beetles devour 3.75 trillion aphids.[5] And think about it: the lady beetles munch only about half the aphids in alfalfa, alfalfa is grown on only 1.0 per cent of California's area, and California is but a tiny fraction of Earth's land mass! There's a mighty mob of ladybugs and aphids rattling around this planet, and when we throw in the other million-plus insect species, the bug numbers become astronomical.

Yes, insects are abundant animals, and they have great genetic plasticity to go along with their numbers. Thus, given species can quickly adapt to unforeseen environmental adversity or opportunity. This is exemplified by the rapid and widespread development of insect resistance to pesticides.[6] Another great attribute of insects is their ability to get around. Most important, they can fly. This means that they can move quickly and for considerable distances, away from adversity or to new food sources. And when they feed, they can do so in a variety of ways, another great advantage. Thus, insects ingest food by chewing, lapping, rasping, sucking, and even absorption through their skins. This enables them to feed on a great variety of foodstuffs under a dazzling array of circumstances.

Indeed, insects are formidable animals, and the key to their success is their superb natural programming to utilize their assets of diversity, adaptability, and prolificity. They don't have to think, for they have built-in mechanisms that take care of that need. And since they don't think, they don't make deliberate mistakes, commit planned atrocities, or try to dominate nature. Their formula for success has worked splendidly for 300 million years, and it is working with great efficiency in their joust with man. On the other hand, we, the thinking animal, have made a basic mistake in assuming that insects, as dumb, lowly brutes, can be easily subdued by the most simplistic of methods. We have ignored the wildly flashing signal of insect success, which clearly warns that the bugs are hard to beat. But, then, we do this repeatedly in our general dealings with nature, which we treat with abuse and contempt.

Let's take a look at ourselves.

Homo sapiens

Our problem is that we are too smart for our own good, and for that matter, the good of the biosphere. The basic problem is that our brain enables us to evaluate, plan, and execute. Thus, while all other creatures are programmed by nature and subject to her whims, we have our own gray computer to motivate, for good or evil, our chemical engine. Indeed, matters have progressed to the point where we attempt to operate independently of nature, challenging her dominance of the biosphere. This is a game we simply cannot win, and in trying we have set in train a series of events that have brought increasing chaos to the planet. There are two major reasons why our challenge to nature was doomed from the start. First, though our brain permits us to plan, create, and execute, its positive traits are overwhelmed by its negative ones. Among living species, we are the only one possessed of arrogance, deliberate stupidity, greed, hate, jealousy, treachery, and the impulse to revenge, all of which may erupt spontaneously or be turned on at will. We simply do not have the capacity to subdue our computer's adverse characteristics, and so our ability to reason, create, and execute is subordinated to the worst of our traits. This leads directly to our second fatal flaw, the lack of a grand strategy to counter nature's simple and successful, age-old game plan. With man there can be no plan, because the human brain is not a single guiding device but occurs instead as a hopelessly unregulated confusion of gray computers distributed among 4 billion people, each computer with its own motivations, ambitions, hopes, hatreds, fears, and jealousies. Result: an ever-deepening earthly tragedy.

It has not always been so. Indeed, for millennia, in our age of innocence, we existed as a creature of nature in near balance with the environment. Then our brain was held in restraint largely by our fear of the unknown. This was the era of tribalism, feudalism, and great religious influence, when nature worked in partnership with the chiefs, lords, gurus, and priests. In that period of innocence and ignorance, nature controlled man just as it does all other creatures, through disease, pestilence, starvation, competition, predation, and climatic or physical catastrophe. And in those days man acknowledged these adversities as the will of God or some other higher powers.

But gradually we began to understand the laws of nature, for as time passed, occasional geniuses appeared among the gray computers and they probed the mysteries of the Earth and the universe. The successive geniuses and their disciples began to piece together a framework of knowledge and technology. Enlightenment emerged from the darkness; the feudalistic stranglehold weakened; the dominance of the religions waned. Suddenly it all came together as a technological and industrial revolution. Overnight, our age-old problems of staying alive, filling our bellies, moving about, and entertaining ourselves became immeasurably easier. We had done it alone, in defiance of nature. We were at the threshold of unbelievable affluence, comfort, health, and enjoyment. Nature was our slave.

Now just look at what we can do to her:

We can hunt the largest mammals with amazing efficiency from vehicles or big ships, with no fear of Simba or Moby Dick. We can pinpoint the breeding grounds of our ancient nemesis the locust and spray the enemy into submission before it forms its ravaging swarms. We can dam the mightiest rivers, control their destructive floods, and use the trapped waters to irrigate our lands and energize our homes and factories. We can prevent or subdue such age-old diseases as malaria, cholera, plague, smallpox, and polio, and can work a thousand other miracles to save and protect humanity from pestilence. We have developed machines that can transport us by the hundreds from continent to continent in a matter of hours. And on a breath-taking scale, we can plow land, pump oil, mine coal, cut timber, fill swamps, level hills, bore through mountains, dig ditches, lay pavement, erect buildings, produce television sets, and crank out "Big Macs."

It all seems so good, but there is a price to pay.

Nature

Nature is tough, and she is showing signs of impatience with our tampering. Think about what's happening. A poorly conceived and engineered dam suddenly collapses and releases a wall of water that destroys a fertile valley that took millennia to evolve. Another dam, developed to bring prosperity to an underdeveloped nation, simultaneously creates a massive breeding place for disease-transmitting organisms, which trigger devastating epidemics. Meanwhile, the suspended silt of the river water is deposited behind the dam, inevitably destroying the reservoir, while at the mouth of the river a fishery dies for lack of river-borne nutrients, and millions face a protein shortage. In another area, men seeking instant wealth greedily strip the forests from the surrounding mountainsides, but with the coming of the next monsoon, a rubble of mud and boulders cascades into the valleys and kills the fertile soils that for centuries had sustained the spoilers' ancestors. In an arid area of the subtropics where nomadic tribes and their flocks had coexisted with nature for centuries, the balance is destroyed when a system of tube wells financed by the aid program of a developed nation leads to excessive production of livestock, which overgraze the land and turn it into a sterile desert. In a great metropolitan center, smog generated by automobile exhaust emissions backs up against the surrounding mountains and destroys the magnificent pine forests that are the recreational outlet for the millions of smog makers.

Yes, Nature is signaling her impatience, and we should heed her warnings and back off from much of what we are doing. But in truth we can't, because our 4 billion gray computers are unmanageable and as such persist in doing their myriad crazy things. To make matters worse, the computers bunch together in groups and factions that play their particular arrogant, greedy, stupid, hateful, jealous, and vengeful games with awful effect. Whites kill Blacks and Blacks kill Whites. Protestants kill Catholics and Catholics kill Protestants. Arabs kill Jews and Jews kill Arabs. Moslems kill Christians and Christians kill Moslems. Communists kill Fascists and Fascists kill Communists. And along with all this killing there is immense abuse of the land and a mindless exploitation of resources.

Brazil says, Don't tell us not to exploit the Amazon jungles

—we've got a lot of environment to spoil. The Japanese and the Russians say, Don't tell us to stop hunting whales—we've got a ton of money invested in our whaling fleets and we want to earn it back. The French-British consortium, and the Russians, with their SST's say, The petroleum shortage, the decibel count, and the ozone shield be damned—we've got billions invested in our SST's and we're going to sell them to regain our investment and enhance our national prestige. The French, Chinese, and Indians say, Too bad, world, we want to join the hydrogen-bomb club and will damned well test our nuclear devices. The forest products industry says, Sure, we're going to spray with DDT—we've got to save these logs and make a profit. The labor unions say, To hell with your wild rivers—we need dams, because dams mean jobs. The petroleum industry says, Forget ecology—we need to rip open the earth to get at the oil shale. The Third World says, Hey, First Worlders, don't tell us not to have babies—that's genocide, and besides we're going to practice all the machismo we want.

Nature sits back and smiles. Her upstart challenger is shot down even before he gets off the ground. She has time on her side, and although she's a bit tattered, she'll be around when the dust settles. And it is a good bet that her favored children, the insects, will be too. In fact, the bug bomb, with our help, is already showing signs of out-megatonning the human population bomb.

Chapter 3

THE PESTICIDE TREADMILL

We do things prodigiously. For example, we gorge on platter-sized, two-inch-thick steaks, drink martinis by the tumblerful, throw away much of the food we place on the table, drive giant, gas-gobbling cars (one for Dad, one for Mom, one for Mike, one for Cindy), maintain our homes at 80° F. in winter and 65° F. in summer, run multiple television sets concurrently, burn lights around the clock in half the house, squander water in bathing, tooth brushing, laundering, flushing, and irrigating, and spray pests in the pattern of saturation bombing. With some of the wasteful things we do, there are as yet no signs of backlash, but with others there are; e.g., a stepped-up incidence of coronary disease, brownouts, and water shortages, and in the case of pest control, a pesticide treadmill.

Now, I am constantly reminded by my colleagues the weed scientists, plant pathologists, nematologists, et al., that *pest* is a generic term and that I am somewhat out of line in using it in reference to insects and their chemical control. They prefer that I use the terminology insecticide treadmill. But these people are really splitting hairs, because what has happened in the form of an insecticide treadmill is also surfacing in the chemical control of weeds, rats, fungi, nematodes, etc. So I do not feel too uncomfortable or technically unsound in using the terms insecticide treadmill and pesticide treadmill interchangeably.

This chapter, then, though largely about insects, insecticides, and the insecticide treadmill, relates in the broadest sense to pest control. So let me tell you how we got onto the pesticide treadmill; in starting out, it is perhaps best to look first at ourselves, the killers.

The Killers

The Grower

A huge, weather-burned man, he shambles out of the stupefying desert heat into the cool, dark womb of the Elks Club bar, swings onto a stool, and roars, "Double Wild Turkey on the rocks." Then, emotionally spent, he sags into his massive frame and absent-mindedly drums his fingers on the mahogany as the bartender crafts the drink. Finally it comes, cool, deliciously mellow, resuscitating. He drains the glass in huge gulps, and the mercy of the booze instantly slams into his guts. Charged, he pounds the stumpy tumbler onto the wood and bellows, "Pablo, another double Turkey." And as the bartender starts up the second magnum dollop, the giant turns to the sorrel-haired, blond-mustachioed strip of rawhide seated to his right and rumbles, "Brad, we really blasted 'em. We really busted the bastards. First we hit 'em with two pounds of methyl and then mopped up with Big Daddy. I swear there ain't a bollworm left on the whole damned ranch."

The rawhide says nothing but slowly nods approval as the respect bordering on love that macho men hold for each other glitters in the ice of his eyes.

The Forester

For long minutes, he bends over his maps in the light cast by a guttering Coleman lantern while a half circle of khaki-clad, hard-hatted men wait silently at the edge of the darkness. Beyond them, silhouettes of helicopters, tanker trucks, and stacked steel drums loom out of the half-murk. Finally, with a cackle of satisfaction rattling in his gorge, the planner straightens up, turns to his unit commanders, and growls, "It's perfect; everything is accounted for. It can't fail. Jump-off time is 0500 hours, and remember, nothing is to remain alive. Nothing! Tomorrow we blast the bloody budworm out of the Modoc."

The Home Gardener

In a half crouch, spraddle-legged, a scowl contorting his face, he confronts the rosebush and spray-blasts it with all the deadly intensity of a frontier marshal pumping the lethal load of his Colt .44 into the thrashing body of a punk gunslinger.

The Housewife

As beautiful and virulent as a coiled fer-de-lance, she waits in a shaded corner of the patio, cold eyes tracking the small winged creature as it moves toward the ambush. Now it is within range, and the killer strikes with blinding speed, hissing out her deadly venom. The stricken animal, an innocuous hover fly, plummets to the ground, spinning out its life in crazy circlings on the flagstones. A cruel smile fleetingly mars the assassin's lovely lips, then fades as she puts aside the aerosol, enters the house, and dials the country club to arrange next week's bridge bash.

Forgive this drift into fantasy; it's the best way I know to dramatize the modern-day approach to pest control. But how did we get into this trigger-happy condition? What turned us on to our killer kick?

The answer is DDT.

Let's take a look at this modern chemical miracle.

DDT

The insects have plagued mankind since antiquity. Of course, in ancient times much of their impact and thievery went unnoticed, because, as hunters and gatherers, we grew no crops to be raided, and in our ignorance of microbial infection, we were totally unaware of the insectan role in disease transmission. But as time passed and we settled into a largely agricultural life style, crop protection became a major concern. Much later, as we developed an understanding of the role of insects in disease transmission, control of the microbe carriers did also. Finally, as standards of living zoomed in the industrialized nations, insect-control efforts were expanded to include species attacking forests, dwellings, clothing, gardens, parks, livestock, pets, and even those that were simply annoying or repugnant. But despite these concerns and the effort to combat a widening spectrum of species, insect control remained, until very recently, a low-grade technology. In fact, we simply accepted the losses inflicted by many of the most serious depredators. Then, in a flash, there was an apparent miracle that promised to end all this nonsense: DDT, a chemical that had been sitting on a shelf for decades, was found to be the most potent insecticide ever tested.

With the discovery of DDT's insecticidal capacity, all traditional insect suppression tactics other than chemical control were shoved to a back burner, as we went for the kill with our new miracle weapon. As an insect killer, DDT worked like nothing ever had before. It killed with rapid and deadly efficiency, was broadly toxic and long-lasting, and it was cheap. It indeed seemed to be the ideal insecticide, a miraculous product of modern technology and the long-sought answer to the bug problem. Why bother with a multiplicity of insect-control tactics! We now had the nearly perfect killer, which by itself could handle most of our tiny tormenters, and what's more, we had the insight to develop additional miraculous materials to handle any problems not solved by DDT. Overnight, pest control was transformed from a side show to the center stage of modern technology. And this technology was overwhelmingly chemical.

DDT catalyzed an explosive expansion of the pesticide industry. The process was simple: The unprecedented effectiveness of this insecticide and its siblings and successors and their low cost evoked an enormous demand for them. This in turn attracted vast amounts of capital to create the production capacity required to satisfy this demand. The resultant expansion of the pesticide industry was so rapid and massive that it simply steam-rollered pest-control technology. Entomologists and other pest-control specialists were sucked into the vortex, and for a couple of decades became so engrossed in developing, producing, and assessing the new pesticides that they forgot that pest control is essentially an ecological matter. Thus, virtually an entire generation of researchers and teachers came to equate pest management with chemical control. So did the grower, forester, food processor, mosquito abater, home gardener, politician, government pest-control bureaucrat, experiment-station director, farm adviser, and just about everyone else directly or indirectly concerned with pest suppression. Their misconception, more than anything else, is what flaws modern pest control. In ignoring the ecological nature of pest control and in attempting to dominate insects with a simplistic chemical control strategy, we played directly into the strength of those formidable adversaries. As a result, today, only a third of a century after the discovery of DDT's insect-killing powers and despite the subsequent development of scores of potent poisons, the bugs are doing better than ever, and much of insect control is a shambles.

The Insecticide Treadmill

There are at least a million insect species on planet earth, but of this vast number only a few thousand—by various estimates, from five to fifteen thousand—have become pests. Of course, many species are innocuous, and most are probably beneficial, acting as pollinators, reducers, scavengers, natural enemies of pests, even human food. Nevertheless, it is probably safe to assume that at least 10 per cent of the insect species are our potential competitors. It would seem, then, that perhaps fifty to one hundred fifty thousand species have a pestiferous potential, yet only a fraction of these ever attain pest status.

Why?

The answer of course is *natural control*, the combination of physical and biological factors in the environment that maintain all species populations within characteristic limits. In other words, there *is* a balance of nature going on around us all the time, and the most broadly affected group of organisms are the insects and their cousins the mites, the earth's most diverse bundle of animal species.

Now, among the biological factors that impinge upon the insects, two kinds of natural enemies—the predators and the parasites—play an immensely important role. Without the ever-active, naturally occurring biological control effected by predators and parasites, it is doubtful that man could stand up to insect competition. What is most amazing about this natural restraint on insect populations is that much if not most of it results from the impact of bug upon bug. In other words, the insects are their own worst enemies, for many of the most important parasites and predators that restrain insect populations are themselves insects with special adaptations for carnivory and a particular taste for bug meat. This is a crucial factor in a burgeoning world-wide insecticide treadmill, which has brought the prevailing, chemical control strategy to the brink of chaos.

Modern insecticides are *biocides*. That is, by design, they kill a wide spectrum of animals. This is the root cause of the insecticide treadmill, for the chemicals kill good bugs as well as bad ones. Thus, if not intelligently employed, they can trigger a bug backlash by interfering with the balance of nature which occurs even in our most severe crop monocultures. For example, when applied to a crop, a biocide

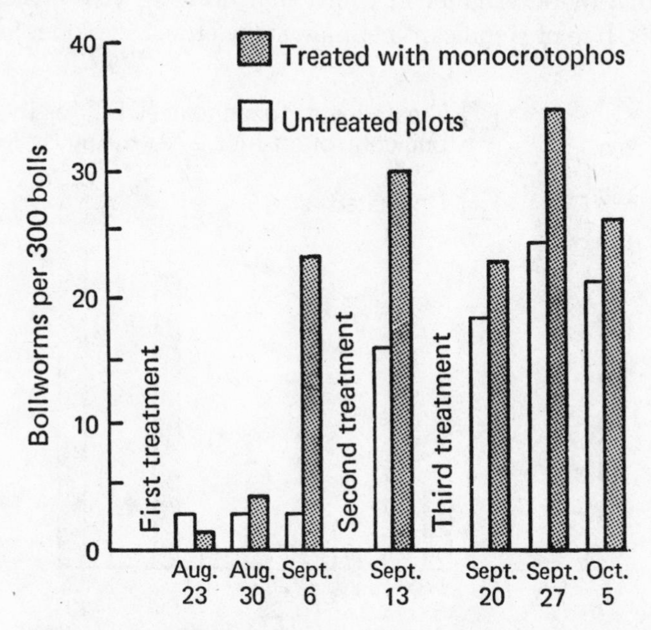

Figure 1. Target-pest resurgence following applications of a "control" insecticide. In this experiment, plots treated with monocrotophos, an insecticide federally registered for bollworm control, suffered heavier infestations than untreated plots. Note particularly the strong bollworm resurgence following the initial treatment. Simultaneous samplings of predators revealed that the insecticide destroyed bollworm predators and thus permitted resurgence of the pest. The data are from an experiment conducted at Dos Palos, California, in 1965. Adapted from R. van den Bosch and P. S. Messenger, *Biological Control*, p. 123 (Scranton, Pa.: International Textbook Co., 1973).

kills not only pests but also other species in the insect community, including the natural enemies that restrain noxious species. Often, the natural enemies suffer excessively, first because they are generally less robust than the pest species, and second, because the insecticides deplete their food supply (i.e., the pest species) so that they starve or leave the fields. As a result, insecticide spraying frequently creates a virtual biotic vacuum in which the surviving or reinvading pests, free of significant natural-enemy attack, explode. Such

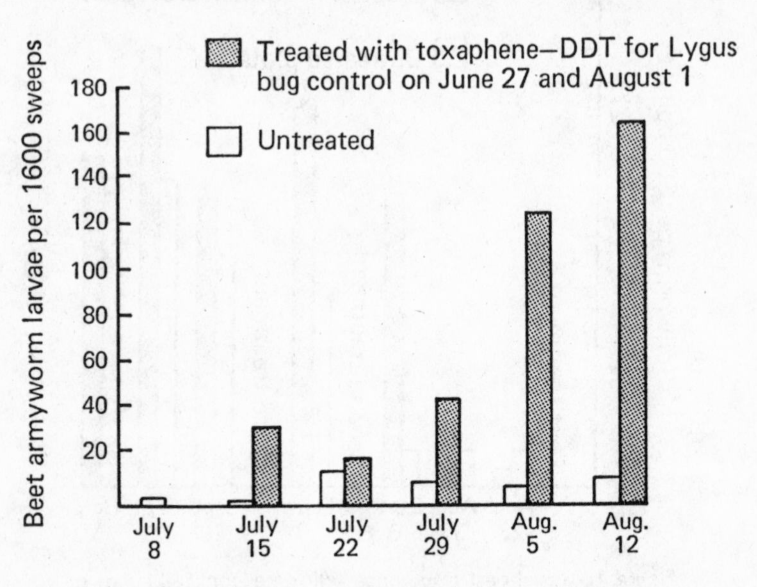

Figure 2. A secondary outbreak of the beet army worm in cotton following insecticide treatment. Note that on August 12 there were approximately seventeen times as many army worms in plots that had been previously treated with insecticide (a toxaphene-DDT mixture) as in plots with no history of treatment. Data are from an experiment conducted near Corcoran, California, in 1969. Subsequent studies have clearly shown that elimination of predators in the treated cotton permits such secondary pest explosions. Adapted from R. van den Bosch, and P. S. Messenger, *Biological Control*, p. 125 (Scranton, Pa.: International Textbook Co., 1973).

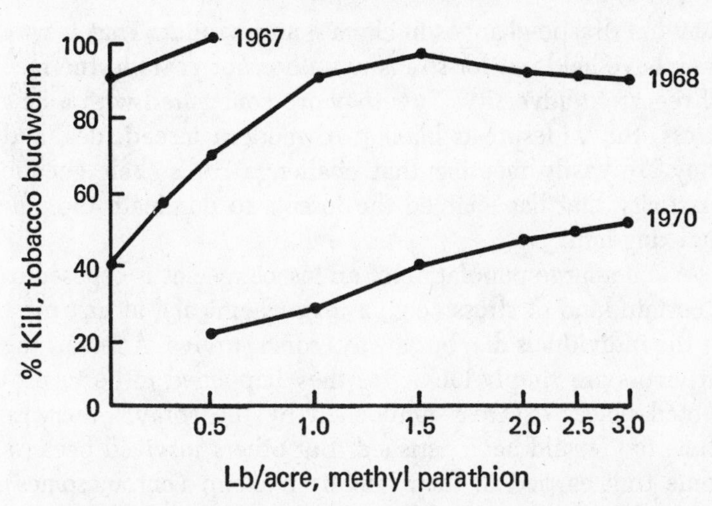

Figure 3. Insecticide resistance. Decline in toxicity of methyl parathion to tobacco budworms from Central Texas during the period 1967–70. Data from P. L. Adkisson, Texas A&M University. Adapted from R. van den Bosch, and P. S. Messenger, *Biological Control*, p. 126 (Scranton, Pa.: International Textbook Co., 1973).

post-spraying pest explosions are often double-barreled, in that they involve not only the resurgence of target pests but also the eruption of previously minor species, which had been fully suppressed by natural enemies. The frequent outcome is a raging multiple-pest outbreak, more damaging than that for which the original pest-control measure was undertaken. Predictably, the grower or other insecticide user, in order to salvage his cotton, fir trees, rosebuds, or whatever, reapplies insecticides, and when this triggers still another multipest outbreak, he sprays again. This is the genesis of the insecticide treadmill, an addictive process that is magnified and prolonged by genetic selection for insecticide resistance in the repeatedly treated pests.

Development of resistance to stress in populations of organisms is a common evolutionary process. The insects in particular have shown a remarkable ability to adapt to sudden and drastic changes in climate and habitat. That is why they have held out for so many millions of years in the face of recurrent adversity. Now they are confronted with a new stress, the widespread blanket of modern insecticides, and they are easily meeting that challenge. It is their genetic plasticity that has enabled the insects to dominate the animal kingdom.

When a large population of an insect species is exposed to a certain kind of stress (e.g., a toxic chemical), at first most of the individuals die, but always some survive. A few of the survivors are simply lucky, for they happened to be in protected places or areas untouched by the spray; otherwise they, too, would have perished. But others survived because some trait carried in their genetic makeup (chromosomes) made them less susceptible to the poison. For example, they may have had the capacity to manufacture an enzyme that detoxifies the poison, or they may have had an integument that prevents penetration of the toxic molecules. Others may have had behavioral patterns that permitted them to avoid the material, still others may have had super-hairy feet, which enabled them to walk over the stuff without touching it.

As the spray process is repeated, the lucky but susceptible escapees of the earlier dosing are sooner or later picked off, but the bugs with traits for survival hang on, and they come to dominate the population, breeding among themselves and producing progeny that are also survivors. It is in this way that large populations of insects become resistant to insecticides, and the more intensive and widespread the poisonous blanket the more rapid the selection for resistance in the pests.

With insecticide resistance plugged into the formula, the treadmill whirs at full tilt, and its consequences can be awesome. The typical outcome of the insecticide jag is crop or resource loss, spiraling control costs, and increased environmental pollution. In its extreme economic impact, the treadmill can cause individual bankruptcies among growers or even the destruction of an entire industry, as happened with cotton in northeastern Mexico.[7] It can also have dreadful impact on the human population and on wild things in the environment.[8]

A look into pest control in California gives insight into the severity of the insecticide treadmill and its economic impact. A neat measuring stick for this is provided by an analysis of the twenty-five most serious pests listed by the California State Department of Food and Agriculture's report on insect-caused crop losses and control costs for 1970.[9] Each of these twenty-five insect species cost the California agri-economy $1 million or more during the 1970 crop year. Their economic impact notwithstanding, the most startling thing about these twenty-five heavy-weight damagers is that 72 per cent of them are resistant to one or more insecticides and 96 per cent are either pesticide-created or pesticide-aggravated.[10] This is a shocking state of affairs but really not surprising, since California is the world's most intensive pesticide user, receiving 5 per cent of the global insecticide load.[11]

What this means is that California agriculture is on a giant pesticide kick for which it and its associated environment are paying an enormous economic, ecological, and sociological price. The economic cost of California's pesticide addiction is strikingly illustrated by the Food and Agriculture Department report just cited, which shows that insect-related costs and losses to the state's crops jumped a staggering 150 per cent between 1970 and 1974; that is, from $254.6 million to $643.3 million.[12] Though inflation played a role in this increase, it hardly accounts for all of it.

What is going on in California is true for the nation as a whole—a massive pesticide orgy in which expenditure, waste, and pollution spiral while pest-control efficiency dwindles. For example, thirty years ago, at the outset of the synthetic-insecticide era, when the nation used roughly 50 million pounds of insecticides, the insects destroyed about 7 per cent of our preharvest crops; today, under a 600-million-pound insecticide load, we are losing 13 per cent of our preharvest yield to the rampaging insects.[13] In other words, a major "reward" of our eleven-fold increase in insecticide use has been a doubling of the bug problem. This reflects incredibly bad technology and extremely poor economics—unless, of course, one is selling insecticides.

Losses and expenditures are only a part of the price we pay for insecticide overuse. Social and ecological costs, though very difficult to measure, are of equal or perhaps even greater importance. What price do we put on a human life lost to parathion poisoning, the massive losses of crop-pollinating insects, the insecticide-induced collapse of an area's economy, or a couple thousand ducks killed overnight by an organophosphate?

The Pest-control Status Quo

The defenders of the pest-control status quo maintain that the case against prevailing insect control practice is a weak one that is completely overshadowed by the good that these chemicals bring. They dismiss the expressions of concern over excess pesticide use and pesticide impact on the environment as a conspiracy against scientific agriculture and agri-business by a coalition of hippies, eco-freaks, organic-food fanatics, and enemies of free enterprise.[14] This purported conspiracy is used by the chemical protagonists as a major ploy in their game to keep things as they are in pesticide and fertilizer use. The thrust of their argument is that, if successful, the irresponsible eco-freaks and their fellow travelers will strip us of our life-giving chemical tools and expose us to devastating pest epidemics, mass starvation, and economic ruin. Heavy play is given to the great benefits that agri-chemicals have brought to humanity, and whenever these materials are criticized, the holy trinity of saved lives, filled bellies, and fat profits is invoked to counteract the criticism. In doing this, the pesticide defenders invariably minimize or ignore the serious problems that attend the use of agri-chemicals, and simultaneously relegate critics of the status quo to the junk heap of irresponsible environmentalism.

Among the pesticide apologists there is a cliché that goes as follows: "If properly used, according to instructions on the label, there is absolutely no hazard in the use of pesticides." Unfortunately, there are major flaws in this cliché. First, the materials are frequently used in ways that deviate from the label, and second, the data that support pesticide registration and labeling are at times inadequate or sloppily obtained and apparently even doctored.[15] Finally, registration and labeling simply do not cover the full spectrum of contingencies, some of which appear only as tragedies in the field.

What else explains the repeated occurrence of unfortunate episodes associated with the use of properly registered and labeled pesticides? Just look at what's been happening in recent times:

—In the autumn of 1971, in a single incident, leptophos, a U.S.-produced insecticide, poisoned a number of Egyptian peasants and killed about twelve hundred water buffalo.[16] Later, in the United States this same insecticide caused permanent nerve injury to workers in the chemical plant in which it was being produced.[17]

—In vast forest areas of eastern Canada and the northeastern United States, nature's silviculturist the spruce budworm, which once functioned to prune out over-aged balsam fir trees at approximately quarter-century intervals, is now perennially epidemic, because chemical control practices designed to save every tree have thrown the forest into a continuously vulnerable condition. The spruce budworm problem is taking on nightmarish dimensions as each year's caterpillars produce masses of moths that spread out over new areas of forest to produce additional destructive populations. Today, 120 million acres of Canadian forest and immense areas of Maine's woodlands are economically threatened by budworms generated out of pesticide use.[18]

—In Central America each year thousands of peasants are poisoned by insecticides used for cotton insect control, and those who survive run the deadly course again and again. Certain of the culpable chemicals are properly registered and labeled, U.S.-produced materials.[19]

—In Mexico in 1973, in an incident in the state of Coahuila, a number of farm workers were killed and hundreds more sickened by several properly registered and labeled insecticides of U.S. manufacture.[20]

—In 1975 in California more than thirteen hundred persons were sufficiently poisoned by pesticides to require medical attention.[21] Since strict laws pertain to the use of pesticides, it is assumed that the victims were poisoned by properly registered and labeled materials. As disturbing as the official poisoning figures are, there is reason to believe that they are conservative, for it appears that many affected persons do not seek medical attention. For example, Dr. Ephraim Kahn, chief of the Epidemiological Studies Laboratory of the California Department of Public Health, estimates that the reported pesticide injuries represent only about 1 per cent of the total poisonings.[22] In other words, if Dr. Kahn's estimate is correct, the actual poisonings in 1975 exceeded one hundred thousand, a shocking state of affairs indeed!

—Today in California the encephalitis mosquito is widely resistant to virtually all conventional larvicides. Resistance in this dangerous species is the result of intensive exposure to insecticides both through direct treatment and indirectly from agricultural applications. Whatever the cause, because of resistance induced by excessive use of properly labeled and registered insecticides, residents of California's Central Valley are today living with a dangerous disease transmitter that is difficult to control.[23]

—In Asia, Africa, and Latin America, malaria, once felt to be on the verge of eradication by DDT, is making a cruel resurgence. Increasing mosquito resistance to DDT, substantially engendered by excessive agricultural use of the material, is a major cause. But this is only part of the story. Another important factor is that DDT never did eradicate malaria in its areas of strong endemicity. The disease was always there, and when the poor Third World countries reduced their spraying programs to divert badly needed funds to other socioeconomic programs, malaria returned with a vengeance among populations that had lost much of their conditioned resistance to the disease. DDT resistance in the mosquitoes, of course, adds an alarming dimension to the problem. For example, in 1976 in Pakistan, twenty-nine hundred persons spraying the supposedly safe substitute material malathion, were poisoned by this insecticide, and five died.[24] What is most disturbing about the mosquito mess is that during all the years when chemical eradication was being attempted, at great expense, other aspects of mosquito control were often neglected.[25]

—In the southwestern United States in the late 1960s a properly registered and labeled organophosphate insecticide caused massive unanticipated bird kills.[26] Then, after a series of "tests" and corrective steps, assurances were given that the material could be safely used. But in 1972 it was again involved in a major bird kill, this time in Florida.[27]

—During one night in the spring of 1974 at Hemet, California, 2,450 ducks resting near a reservoir were killed by still another properly registered and labeled insecticide that apparently drifted over them during the spraying of a nearby alfalfa field.[28]

—In California in the spring of 1972 approximately thirty-five thousand tons of alfalfa, valued at $1.5 million, was declared unfit for marketing due to contamination by residues of a properly registered and labeled insecticide used for weevil control.[29] In 1975 in a similar episode a second properly registered and labeled insecticide spoiled an additional several thousand tons of alfalfa hay.[30]

—In August 1974 a Yuma, Arizona, aircraft spray firm contaminated the Yuma main canal with a mixture of properly registered and labeled pesticides and other agricultural chemicals, killing numerous fish and forcing closure of the Yuma water works.[31]

—Each year in the U. S. Southwest, tens of thousands of honeybee colonies are destroyed by properly registered and labeled insecticides.[32] For example, on the western side of California's San Joaquin Valley, pesticides destroy 25–30 per cent of the colonies each year. The problem became so critical that the apiarists were threatened with financial disaster. But honeybees as pollinators are vital to agriculture, and so the agri-business lobbyists prevailed upon Congress to pass an indemnification law to compensate the beekeepers for their losses. As a result, the beleaguered taxpayer pays the bill so that the agri-chemical industry can keep on dumping excess amounts of its bee-killing poisons into the environment. But now the apiarists have found that the indemnity is not sufficient to cover their losses, because it pertains only to colonies that are directly destroyed. Pesticide-weakened colonies, which slowly fade away, are not covered.[33]

—In California's San Joaquin Valley in recent years many small farmers have been pushed down the road to bankruptcy by pest outbreaks (e.g., cotton bollworm) triggered by unnecessary treatments of properly registered and labeled pesticides foisted on them by agri-chemical salesmen.

When the farmer takes his case to court, he invariably loses, because the high-powered agri-chemical-company lawyers convince the judge or jury that the bug outbreak was an act of God or the plaintiff was an inept farmer.[34]

—Globally, spider mites, once relatively minor crop pests, have been boosted to the top of the pest heap by the properly registered and labeled insecticides that were supposed to solve our insect problems. This has come about largely because the pesticides suppress the spider mite's natural enemies. The materials also have at times physiological effects that stimulate spider mite reproduction.[35] Today in California the once relatively unimportant spider mites cost the agri-economy more than $116 million annually, double the losses caused by any other "insect" pest group and five times what they cost the economy fifteen years ago.[36]

—In 1975, workers in a Virginia chemical factory producing Kepone®, a properly registered and labeled organochlorine insecticide used in roach and ant poisons, developed various hideous symptoms of insecticide poisoning.[37] The victims were poisoned apparently because they came into direct contact with Kepone® under the lax operating conditions that prevailed in the chemical plant. Because of the poisonings, the plant was closed down and dismantled. But then it was found that the surrounding countryside, including the James River, was also contaminated, forcing closure of parts of the river to fishing. Meanwhile, what is society going to do for the unfortunate, permanently affected victims?

—In the 1960s, DDT and other properly registered and labeled organochlorine insecticides virtually eliminated the brown pelican from the Gulf coast of Louisiana. This majestic creature is Louisiana's state bird, and the proud Bayou Staters could not bear the thought of its disappearance, so they arranged to import a vigorous new breeding stock of pelicans from Florida. But in 1974 virtually all of the imported birds and their progeny were wiped out by endrin, a properly registered and labeled insecticide that was flushed by floodwaters into the Gulf of Mexico from Louisiana's cotton fields.[38]

The above listing is just a sample of the economic, social, and ecological problems associated with prevailing pest-control practice. And since virtually all of these incidents have involved properly registered and labeled insecticides, their consequences make a hollow mockery of the "follow the label" cliché.

The most disturbing cases are those that involved "unforeseen" impact, such as the bird losses in the Southwest, the Kepone® and leptophos tragedies, and of course, California's DBCP tragedy. I find it extremely odd that in each of these situations, somewhere along the line of research and development someone should not have discovered the hidden hazards that these pesticides posed and warned of those hazards. Whatever the reasons, the history of tragedy that has just been chronicled should make it abundantly clear that in playing fast and loose with its pest-control biocides, society is really indulging in environmental Russian roulette. Quite frankly, pest control is in a state of chaos, and the associated problems have placed a staggering impost on society and the environment. In this connection two recent revelations add dimension to this chaos. The first is an editoral in the prestigious journal *Science* (January 27, 1978), reporting that our agricultural production is beginning to plateau. In other words, despite intensifying agro-technology inputs, our crop plants are approaching their genetic limits of production, and there is simply not much more that we can squeeze out of them. The second revelation is that despite this flattening out of crop production, pesticide use is increasing exponentially. In fact, as world-renowned entomologist Robert Metcalf stated in a deposition to a hearing of the California Department of Food and Agriculture

(Sacramento, December 20, 1977), the doubling time for the volume of pesticides used is eight years. That is to say, by 1986 our environment will suffer double the pesticide load it receives today, while crop yields per acre will at best increase only slightly.

Good business for the biocide hustlers, but appalling news for the user and the biosphere!

Sixteen years ago, *Silent Spring* awakened the world to the problem of environmental pollution and warned particularly of the dreadful threat posed by pesticides. For a while, there was general concern about pesticide impact on wildlife and the threat of pesticide poisoning and pesticide-induced carcinogenesis, teratogenesis, and mutagenesis to mankind. But Rachel Carson's was an isolated voice in the vast human chorus, and though clear and sweet and strong while it lasted, it was mortal, and it was abruptly silenced by death.

Other voices, more diverse and powerful, survived, and they quickly muffled the echoes of *Silent Spring*. And so we slipped back to our old ways. Today the pesticide treadmill spins more wildly than ever. We use twice as much insecticide as in 1962, there are more insect species of pest status than ever before, insect control costs have skyrocketed, and insecticide impact on the environment is the worst in history.

Homo sapiens, the only deliberate fool that ever evolved, is back tending shop in the good old way!

Chapter 4

THE MELANCHOLY ADDICTION OF OL' KING COTTON*

Nowhere is the pesticide treadmill so devastatingly manifest as in cotton.[39] It is for this reason that I have chosen the cotton story to serve as a model for what has so often gone wrong with insect control in many of our heavily sprayed crops.

Cotton today is one of the world's most "bugged" crops, victimized by an ecological backlash to heavy insecticide drenching. The sad state of the cotton ecosystem stands out as an example of the worst in pest control. The heavy use of pesticides has created an entomological nightmare, bringing in its wake economic ruin, human illness and death, and gross environmental pollution.

Why cotton? How was this ecosystem, this crop, programmed for disaster? First, over its history cotton has been plagued by a variety of destructive pest insects that require control. Second, as a cash earner, both for the individual grower and for governments seeking foreign exchange, cotton often becomes a political crop. Governments play with it: set acreage allotments, fix prices, pay cash subsidies to growers, make barter arrangements with other countries.

* This chapter is an updated version of an article published in *Natural History* magazine in December 1968.

Because of these factors, wherever cotton is grown farmers use insecticides almost compulsively. The sight or even thought of a boll weevil, lygus bug, spider mite, or pink bollworm triggers an automatic reaction: kill it before it grubs a nickel out of your pocket or a crumb from your mouth, or before it milks the treasury of a single peso! Many growers, whether on their own resources or financed by banks or ginning companies, use insecticides prophylactically, often because of pressures from the lending agencies.

In the United States in past years the government cotton subsidy system also fostered excessive use of insecticides. Subsidies were based on average yield records: the higher the average over a prescribed number of years the greater the subsidy. The grower, believing "insect-free" cotton to be a critical element for maximum yield, often used insecticides prodigally, regardless of pest infestation levels. To him the cost of insecticides was insignificant compared to the potential earnings from subsidies. Even today, with subsidies abolished, many growers, conditioned by past practice, continue to strive for bug-free cotton.

Governments, too, become extremely nervous over cotton insects. A number of countries that barter cotton for manufactured goods or earn substantial foreign exchange from cotton exports are obsessed with the same urge as individuals to maximize the crop by protecting it from insects. Again, this means heavy insecticide use.

Finally, there is the influence of the pesticide industry itself. Cotton growers are the greatest insecticide buyers, and the chemical producers compete fiercely for a piece of the action. In the United States alone, cotton insecticide sales exceed $100 million annually. There are other great markets in Latin America, Africa, Australia, the Middle East, and Asia.

Many pesticide companies play on the apprehensions of growers and governments, bombarding them with advertisements and "free" technical advice. Often these companies give grants and provide experimental materials to research institutions. The institutional staff members are entertained, showered with little gifts, and even given funds to visit chemical production facilities and research projects and to attend meetings, or just to take a holiday. Individual growers and governments are offered price discounts on proprietary materials. And in some instances, agriculture officials are influenced to approve the purchase or use of specific materials (an embittered insecticide-company representative once related to me in detail how, at the last minute he lost a contract with the Iranian Ministry of Agriculture when an agent of a rival company bribed a key ministry official).

If individuals or governments are coaxed, tricked, or bribed into excessive use of insecticides, why not just forget it and let them suffer the consequences of their own stupidity? Unfortunately this is only one side of the coin, for many innocent victims have been caught in the backlash of the insecticide overload. The result has been massive economic, sociological, public-health, and ecological adversity.

The basic problem lies in the ecological crudeness of most modern insecticides and the patterns of their use, which, as was explained earlier, engender pesticide treadmills with severely adverse impacts.

Cotton, as the world's major recipient of the insecticide overload, has suffered from pesticide addiction more than any other crop. Major insecticide-induced breakdowns of cotton ecosystems have occurred in a number of areas: Egypt, South and Central America, Mexico, the Rio Grande Valley of Texas, the Imperial Valley of California, and the Ord River Valley of Australia. The landmark disaster occurred in the Cañete Valley of Peru in the early 1950s.[40]

The Cañete is one of a number of Peruvian coastal valleys that are self-contained agro-ecosystems. Major cotton production in this valley began in the 1920s. Until the late 1940s, cotton growers controlled a limited pest-insect complex with old-fashioned insecticides such as calcium arsenate and nicotine sulphate. Then the growers, opting for modern pest control, shifted to the new, synthetic organic insecticides, mainly DDT, benzene hexachloride (BHC), and toxaphene. During the first years, the modern pesticides effectively killed insect pests, and striking increases in yields were recorded. Because of the success, the growers spread a virtual blanket of insecticides over the valley.

But the miracle was short-lived. Some of the pest species began to develop resistance to the insecticides. Old pests became increasingly destructive and new ones appeared. By 1952, BHC was no longer effective against the cotton aphid, and by 1954, toxaphene failed to control one of the leafworms. Quickly, a general pattern of resistance to organochlorine insecticides developed. The growers turned to the deadly organophosphates. But the pests, whose species numbers had doubled, rapidly developed resistance to these organophosphates, too. The growers had to increase dosages and shorten treatment intervals. By the 1955–56 season, insect resistance to the organophosphates was general, the pests were rampant, and the crop suffered severe losses. The yield for the 1955–56 season was one of the lowest ever recorded in the valley.

To rescue the cotton industry, the Peruvian growers broke away from their overdependence on insecticides and invoked an integrated control program that, among other things, included legal restrictions on use of the synthetic organic insecticides, a reversion to some of the old, nonsynthetic materials, adjustments in agronomic practices, and rehabilitation of the natural enemy fauna. The pest problem abated rapidly. Secondary pests, which had been triggered to damaging abundance by the synthetic insecticides, faded into obscurity. The old regulars subsided to their former levels. Damage decreased and yields jumped. Under integrated control, Cañete Valley cotton production quickly reached its highest levels in history and has remained there.

The Cañete Valley disaster had its sequel in Central America. The problem there, however, was of greater magnitude and had serious effects outside the cotton fields. The pattern of target pest resurgence, secondary pest outbreak, and pesticide resistance developed in hundreds of thousands of acres of cotton, extending over much of the Pacific plain of Central America. Production costs soared and yields dropped. The yield losses had a particularly severe economic effect, because several of the Central American countries depend heavily on cotton for foreign-exchange earnings.

But the economic woes of Central American cotton were only part of the insecticide-induced disaster. The direct impact of the materials on the human population was appalling.[41] In Nicaragua alone, there were 383 reported deaths and over three thousand poisonings during the 1969–70 crop year. A study in Guatemala showed that DDT and BHC residues in mother's milk were the highest ever recorded anywhere. In an extreme case, a sample of mother's milk contained 244 times as much DDT as would be permitted in commercial cow's milk in the United States.[42]

Another public health menace associated with the overuse of insecticides in Central America's cotton was the increased threat of a malaria epidemic due to insecticide resistance in the non-target, malaria-carrying mosquito *Anopheles albimanus*. This resistance resulted substantially from the veritable chemical cloud that recurrently covers much of the countryside during the cotton-growing season. Recent studies in Central America indicate that malaria is on the increase there.[43]

An indirect public health threat has also developed from the cotton mess. This involves a species of whitefly and a leafhopper generated as secondary-outbreak pests in the wake of cotton spraying. The whitefly and leafhopper are induced to outbreak status by pesticide drift from treated cotton onto beans and corn, resulting in reduced production of these staple foods.[44] The adverse impact on beans, a key source of protein, is particularly disturbing.

Central American politicians, agriculturalists, and public health officials have recognized the source of the cotton pest control problem and have been seeking an answer to it. For example, in Nicaragua an integrated control program developed jointly by Nicaraguan specialists and United Nations (FAO) entomologists has resulted in a substantial reduction in pesticide use and an increase in yield.[45] Unfortunately, prevailing socioeconomic patterns and political corruption offer formidable obstacles to a general reduction of the pesticide overload in Central American cotton. In this light, it can only be hoped that an area-wide program on the pattern of that developed in Nicaragua will evolve. Meanwhile it is saddening to realize that so many have had to suffer and continue to suffer because of a pest-control system developed largely out of greed.

In two areas of northeastern Mexico, the "right action" did not come soon enough; in both places the pesticide treadmill destroyed the cotton industry. In this case, a single species, the tobacco budworm, largely caused the disaster. Formerly controlled by natural enemies, the tobacco budworm was freed from them by insecticides used to control the boll weevil and cotton fleahopper. The inherently tough tobacco budworm became tougher as repeated exposure to insecticides eliminated the weaklings from its population. Eventually, no insecticide dosage or combination could kill it. Despite repeated drenchings, the insect munched three quarters of a million acres of cotton at Matamoros-Reynosa into extinction and later did the same thing to a half million acres at Tampico-Ciudad Mante. The cotton crop at Matamoros-Reynosa once had an annual value of about $50 million, that at Tampico-Ciudad Mante about $33 million. Now these once prosperous regions are experiencing a severe economic depression.[46]

Across the Rio Grande, in Texas, symptoms of a similar debacle appeared. The tobacco budworm became increasingly difficult to kill, damage caused by it increased, and pest-control costs soared. But in the nick of time the Texans averted disaster by minimizing the early-season treatments for the boll weevil and cotton fleahopper, which trigger the budworm outbreaks.[47] Nevertheless, in years of heavy boll weevil or fleahopper infestations, extensive insecticide treatment is necessary, and as happened in 1970, this engenders a severe tobacco budworm problem. Somehow, the early-season treatments must be permanently reduced, but there is no easy solution to this enigma. The Texans are managing to live with the problem while they are developing an integrated control program. Recent developments in this program have been extremely promising.

The cotton industry in the Imperial Valley of California has experienced much the same economic and ecological distress as has that of the Rio Grande Valley. First, an invading pest, the pink bollworm, almost ruined the industry outright. The pest invaded the Imperial Valley, presumably from Arizona and Mexico, in the mid-1960s. The lush valley, with its torrid summers and mild winters, seems to be an ideal environment for the pink bollworm, for the pest has prospered there as it has nowhere else. Of course, human bungling has helped it too.

When the pink bollworm arrived in the Imperial Valley, cotton growers, the U. S. Department of Agriculture, Agricultural Research Service (ARS), the California Department of Food and Agriculture, and politicians—all in an atmosphere of hysteria—made several major mistakes. First they undertook an ill-advised attempt to eradicate the pest with insecticides.[48] This plan was foredoomed. For one thing, insecticides are not particularly effective against the pink bollworm, whose larvae feed cryptically, within the fruit. What's more, contiguous infestations in Arizona and Mexico assured reinfestation of the valley even if by some miracle the chemicals did temporarily eliminate the pest. Not only did the program fail, it also had unfortunate secondary effects: (1) many hundreds of thousands of dollars, which could have been invested in research on integrated control, were squandered, (2) tens of thousands of honeybee colonies were destroyed, (3) devastating secondary pest outbreaks were triggered, and (4) insecticide resistance was accelerated in damaging secondary-outbreak pests such as the cotton leaf perforator.

Next an attempt was made to eradicate the pink bollworm by the sterile-male technique. But the technology for successful application of the technique had not been developed, and of course the moth kept boiling in from Arizona and Mexico. Again a large amount of money was expended on a program that failed.

Meanwhile, the growers, whose self-imposed assessments had been largely plowed into the "eradication" programs, still had the pink bollworm in their fields. They fought it in the only way they knew—with chemicals. And predictably, the familiar pattern unfolded: costs for insecticides soared, secondary pests appeared, resistance developed, yields dropped. Furthermore, the thousands of acres of cotton in the Imperial Valley became an enormous insectary, from which millions of insects fanned out over the countryside, infesting a variety of crops. In fields adjoining cotton, additional pests were unleashed because their biological controls had been disrupted by insecticides that drifted from the treated cotton. Among the secondary-outbreak pests in these non-cotton crops, the main culprit was the beet army worm, which developed devastating infestations in sugar beets and alfalfa and also caused damage to lettuce. Aphids, too, occurred in unprecedented abundance.[49] More recently, the dreaded tobacco budworm has appeared in damaging abundance for the first time.

In 1970, cotton yields in the Imperial Valley were the lowest of the post-World War II era. Pest-control costs were staggering. The growers considered a year's moratorium on cotton production as a possible way to break the pink bollworm cycle and its attendant insecticide treadmill. But agreement on a moratorium was not reached, and the Imperial Valley cotton industry continued its annual insecticide drenchings at enormous economic and ecological cost.

The major hope for cotton in the Imperial Valley lies in the development of a integrated control program based on early maturity of the crop, early harvest, and early destruction of the crop residues in which the pink bollworm larvae overwinter. This would prevent the build-up of heavy and destructive populations in the autumn and the production and survival of large numbers of overwintering (hibernating) larvae.

But a combination of factors has frustrated this development. The first of these was the squandering of potential research funds on the fruitless eradication programs, the next has been grower greed as expressed in their striving for maximum yields by extending the production season into the late autumn, and finally there has been the beguiling sophistry of the agri-chemical industry in its hollow promise of pink bollworm control with miracle insecticides.

The result of all this has been an entomological disaster. The pink bollworm has developed increasing resistance to the insecticides, and now, as chemical treatment has been stepped up, that old nemesis of cotton the tobacco budworm has appeared on the scene in full force. This tough brute, resistant to all available conventional insecticides and freed of its natural enemies, has exploded to such immense abundance that in 1977 it destroyed half the crop—$50 million worth.

I worked the cotton fields of the Imperial Valley for twelve years, and in all those years I never saw a single tobacco budworm. Today it is rampant. The pesticide treadmill has come full circle in the Imperial Valley.

The history of cotton insect control has been marked by waste, misery, death, and destruction. Yet we seem incapable of learning from this pattern of disaster. The growers and entomologists of Central America, northeastern Mexico, and the Rio Grande and Imperial valleys apparently did not profit from the experience of the Cañete Valley. Now in Australia the relatively new cotton industry has repeated the same mistakes. The Australians have spread an insecticidal blanket, and already the treadmill has taken its toll. The situation in the Ord River Valley became so bad that an ambitious cotton-growing scheme there was abandoned. More recently, bollworms have become rampant in the Queensland-New South Wales cotton growing area, and the industry there is threatened with economic chaos.[50]

What fools we are! Insects are our most successful rivals for the earth's bounty, yet when we attempt to suppress them we insist on playing into their strength. As we continue our folly, the repeated triumphs of these little beasts may well be the first faint indicators of our own demise.

Chapter 5

THE MAKING OF AN ECO-RADICAL, OR PARDON MY PARANOIA

"You tell half-truths."[51]

"You are a scientific fraud."[52]

". . . a disgrace to the university."[53]

". . . a sensation-seeking intellectual prostitute."[54]

"One of my entomologist friends suggested that if he were to rate scientific integrity on a 0 to 10 basis you would rank in parts per million (ppm)."[55]

"We don't know how much the upkeep is, but even if he were a dollar a year man, the price is much more than California can afford."[56]

"We suggest that the public look into his background and reasons for his tirade against the free enterprise system. . . ."[57]

"He is a charlatan."[58]

These "endearments," voiced by a spectrum of characters ranging from trade-magazine editors to Berkeley colleagues, are part of the price that I, a research biologist, have paid for joining the public controversy over pesticides. This barrage of invective, though hard to take, has at least helped me to understand why scientists have so little appetite for public debate on technological issues. I don't admire those scientists who lie low when they have the facts to speak out for the public welfare, but I can understand their choice of silence. On the other hand, I have nothing but contempt for those among the silent scientists who, at moments most opportune to them, dart from their safe little burrows to denigrate those of us who elect to speak out on issues. Nothing has saddened me more than the surfacing of several of these people among my immediate colleagues.

What was it, then, that turned me into an eco-radical, willing to opt for such abuse and emotional stress? I shall try to explain.

A long-standing stereotype depicts the entomologist as a preoccupied and rather ineffectual soul who gets his kicks cruising the countryside in purusit of butterflies, beetles, and bees or by probing the sex life of such creatures as tiger moths, stone flies, stinkbugs, and piss ants. To tell the truth, this was pretty much my vision of the future when I became completely hooked on bugs, forty years ago.

Insects have always fascinated me. As a tyke, I constantly relieved the neighborhood gardens of their lady beetles, bumble bees, and butterflies, and cluttered the house with my prizes. Today, my happiest hours as a professional entomologist occur when I am collecting or observing insects in the field. The majority of my friends are entomologists, for the most part gentle, scholarly people who occupy themselves with the biological doings of such creatures as lady beetles, plant lice, fruit flies, mini-wasps, chiggers, wolf spiders, and similar obscure but fascinating species. And if the joy in having so many good friends were not enough, I am additionally blessed in having a most wonderful job, as professor of entomology at the University of California, Berkeley.

It would seem from what I have just said that life for me is, indeed, the carefree bug binge that I envisaged as a youth. But, sadly, this is not so. Instead, the idyllic world of beetles and butterflies has largely slipped away as I have become increasingly involved in the roaring pesticide controversy—a vicious, nerve-wracking imbroglio that has turned my entomological niche into a veritable hornet's nest. What is most saddening is that, as I have become increasingly enmeshed in the pesticide hassle, I have turned into a ruthless gut-fighter in a slugfest without rules or a semblance of fair play. This metamorphosis has been forced upon me because it is the only way that I can hold my own in the shoot-out, or for that matter even survive it.

It is difficult for a scientist to play this way. We are largely a tribe of preoccupied people who just want to be let alone to do our thing. Since we lack the appetite for slashing combat, or the skills to survive connivance, we usually avoid confrontation. But sometimes the things we cherish are threatened, and then we must either take a stand or be overwhelmed in our passivity. Either way, we pay a price, for if we choose to fight effectively, we must make unpleasant character adjustments and divert time, energy, and thought from the things we would rather do. But if we remain indifferent, we stand to lose much of what we love, not to mention our self-respect.

I do a number of things in entomology, but in essence my concern is pest control. I have been in this game since 1946, when I returned from World War II and began my graduate studies at Berkeley. As for the pesticide hassle, I suppose I was programmed right from the start to get into the thick of things. I can put much of the responsibility for this on Berkeley professors Abe Michelbacher and Ray F. Smith, two of the early integrated-control proponents who infected me with their ecological toxins before I had a chance to discover some safe specialty such as insect classification (taxonomy) or conventional pest control. They taught me that insect control is an ecological matter, and already in 1947 I was working in a prototype integrated pest-management scheme: Smith's alfalfa butterfly control program in California's San Joaquin Valley. During that experience, I witnessed one of the early pest outbreaks engendered by a modern insecticide. This occurred when a DDT-induced spider-mite infestation destroyed fifteen hundred acres of alfalfa. It was a lesson I never forgot.

We did a lot of screaming about the dangers of undisciplined pesticide use in those days, but we were only a handful, ahead of our time, and virtually no one paid attention to our rantings. In truth, this was fortunate: nobody tried to squash us. We were evidently deemed harmless. So we lived to scream another day by assuming various guises, I as a biological control specialist.

I have had a very happy and fruitful career in biological control, having been intimately involved in perhaps as many successful programs as anyone of my time. I suppose I should have been satisfied with this—kept my mouth shut and maintained a low profile. But my horizons are broader than biological control, and that is how I got into trouble. My deep involvement in, and increasing understanding of, biological control convinced me that it could best function in integrated control systems. I came to know the true scope of biological control, its attributes and limitations, and the factors that disrupt or augment it. This led me into studies with several colleagues to develop selective insecticide use to complement predators and parasites. For similar reasons, I developed a co-operative relationship with Berkeley colleague Louis Falcon, a specialist in insect microbial control, and I also got involved with University of California, Riverside, colleague Vernon Stern, in cultural manipulations of alfalfa and cotton as they affect pest control. It was my further good fortune to work in the highly successful spotted-alfalfa-aphid integrated control program, which involved use of pest-resistant alfalfa varieties. From these and other experiences I came to see how the pieces might be fitted together, and I became impatient to get on with the game of integrated control.

But while I was having these experiences and developing a deepening conviction of the validity of the integrated control concept, significant things were happening in the pesticide industry, events that conflicted directly with the objectives of integrated control. The worst was the explosive increase in insecticide use. The early successes with DDT triggered the development of an array of similar, organochlorine insecticides, which were then joined by the organophosphates and carbamates. Pesticide use burgeoned, and chemical-plant production soared almost exponentially.

We weren't getting rid of insects, but to many people the chemical panacea always seemed just around the corner, and some folks were having a ball—getting fat off the increasing pesticide market. It wasn't all smooth sailing for them, though. Legislation (the Delaney amendment) and *Silent Spring* took some of the fun out of things. But these were just irritations. Matters were under control. The pesticide proponents, reinforced by the good old boys in the U. S. Department of Agriculture (USDA), the aggie colleges, and the state agricultural-department bureaucracies, made sure that the public understood that the only good bug was a dead bug. And they saw to it that bug killing remained pretty much as it had been, despite Rachel Carson and the eco-freaks. In fact, one of our leading university entomology departments derisively dubbed its pesticide-storage shed Rachel Carson Hall, a mocking testimonial to the myopia of the bug-killing establishment.

But then the bugs began to spoil the act. They just wouldn't roll over and die. In fact, increasing numbers of species rose to pest status. What's more, many of the worst ones developed resistance to pesticides. As a result, control costs soared, there were pest-control breakdowns, and most ominously under the burgeoning chemical blanket, pesticides became major pollutants. This latter development raised the apprehensions of environmentalists, who launched an aggressive anti-pesticide campaign. The pest-control establishment reacted to this by describing the campaign as irresponsible eco-hysteria. They asked, "What do you want, folks, a few lousy pelicans, or millions of filled bellies, saved lives, and bulging pocketbooks! Tell the eco-freaks to bug off."

But in brushing aside the anti-pesticide clamor as a passing irritation, the agri-chemical establishment completely underestimated the tenacity and clout of the environmentalists. The latter came on like aroused wasps. And I chose to fly with them.

Now, how in the world did that happen?

One event more than any other led me into this alliance. This was the Azodrin® affair. Azodrin® is an organophosphate insecticide produced by the Shell Chemical Company. In the middle 1960s, when it appeared on the scene, it was heralded as a highly promising material for cotton bollworm control. At that time we seemingly needed such a material in California, for the bollworm was rampant in cotton and there was no insecticide to control it. The relatively effective DDT had been severely curtailed because of its environmental hazard, and many of our cotton growers (particularly the smaller ones) were in desperate economic straits because of the bollworm assault. (It was only later that we found the bollworm to be an insecticide-induced pest.)

Under these circumstances, I joined university colleagues Louis Falcon and Thomas Leigh and several Shell entomologists in a series of experiments to test the effectiveness of Azodrin® against bollworm. For three summers we worked arduously in a concerted effort to adapt Azodrin® to our growers' needs. But it failed to measure up, and we could not recommend its use. To close things out, at the company's invitation I presented a seminar on our joint Azodrin® study to the Shell research staff. In this lecture I fully discussed the material's shortcomings and the reasons why the University of California could not recommend it. I thought that this took care of the matter, but how naïve I was! Two seasons later approximately one million acres of California cotton were treated with Azodrin®.[59]

Shell seems to have been little impressed by the negative findings of the joint research program. I was stunned. Not because I knew of the shortcomings of Azodrin® and the problems engendered by its use, but because for the first time I realized that I and other university researchers had virtually no influence over pest-control policymaking. It had been rudely brought home to me that over the years we university types had simply been puppets playing silly little games while the pesticide establishment called the shots in pest control. The hope of developing integrated control was a vague dream, and scientific pest control a farce.

Frankly, I was hurt, humiliated, frustrated, and very angry, and when I cooled down I made up my mind to do everything possible to turn things around. But how does a lonely dissident go about doing this? It was perfectly clear that the USDA, the land-grant universities, the professional societies, the grower groups—all the normal channels of action—were of no use, for they were part and parcel of the pro-pesticide establishment, quite happy with the status quo. To attempt to beat a revolutionary drum among them would have been an act of utter futility. I had to look elsewhere, and there was only one way to turn: to the other camp in the pesticide hassle, that of the environmentalists. I homed in on the "bird and bunny lovers" like an ant on a pot of honey, and in doing this my antennae pointed to the June 1968 Toxicology Conference at the University of Rochester.[60]

It is a remarkable fact that in all the conferences, symposia, and colloquia concerning pesticide pollution conducted prior to the Rochester Conference, entomologists played virtually no role. Those meetings characteristically involved insect toxicologists, human toxicologists, residue chemists, public-health specialists, and wildlife biologists, but not economic entomologists. This was just another symptom of the impotence of entomologists in shaping pest-control policy. In other words, we bug chasers were of such humble estate that no one even bothered to seek our participation in sessions concerning the adverse consequences of pest-control programs. We were clearly a low-grade profession, charged with running errands for the prime movers of pest control. It was quite a feat, then, when I wangled an invitation to the Rochester Conference through the efforts of Robert Rudd and his honcho Steve Herman, staunch friends of the peregrine falcon, brown pelican, and western grebe. The Rochester Conference opened a multitude of doors. I had never before encountered such a horde of eco-activists. It was a unique experience to hear a hundred voices simultaneously bad-mouthing prevailing pesticide use. And for their part, the environmentalists finally had a bona fide aggie-college entomologist in their midst who understood their concerns. In happy union, we crashed into each other's arms. The environmentalists needed an entomologist to help them in the pesticide controversy, and I, an entomologist, needed their resources, know-how, and political clout to support my entomo-radicalism.

I had found the vehicle for my message, a vehicle that of necessity has been more weapons carrier than sports car. And though the ride has been mostly one of bumps, bruises, and bombast, it has had its rewarding moments, as when a gentle soul responded to one of my published articles by assuring me that all was well because I was "blessed by God and the angels." Vibes like that keep an old "charlatan" going.

WATER BUGGING

Chapter 6

THE POLITICS OF PEST CONTROL

The word "pest" refers to a wide range of plant and animal species that annoy us, endanger our health, attack our cherished possessions, or rob us of food and fiber. Because pests are noxious, obnoxious, and larcenous, we feel that we must control them, and this we frequently try to do with toxic chemicals.

As an entomologist, I am most familiar with pestiferous insects, and so in this discussion the term "pest" will largely relate to these tiny competitors. However, what is said about politics and bug killing has wide application in the over-all field of pest control and thus truly reflects the politics of pest control.

It should be clear from what I have written so far that much of modern pesticide use is excessively costly, inexcusably inefficient, and shamefully pollutive. Some in the pest-control game are aware of this and have loudly decried the situation for years. But nothing much has come of this, because a very powerful coalition of agencies and individuals who don't want change have successfully muted the cries of dissent and thwarted efforts to effect reform. This pro-pesticide consortium is very comfortable under the prevailing system, wants things to remain as they are, and plays political games to maintain the status quo. Thus, in an era of increasing concern over food production, energy shortages, and environmental quality, politics is helping to perpetuate a costly, inefficient, and pollutive pest-management system.

Who, then, are the members of the consortium that opts for this seemingly undesirable state of affairs? *Topping the list, as one would expect, is the agricultural-chemical industry.* The pack also includes:

1. pest-control operators (e.g., farm service and supply companies, termite exterminators);
2. aircraft applicators (e.g., spray-plane operators and their organizations);
3. agri-business concerns (e.g, banks, utility companies, farm equipment manufacturers);
4. grower organizations (including marketing co-operatives as well as lobbying-type organizations);
5. food processors;
6. certain key politicians (particularly those from the corn and cotton belts);
7. administrators, elements, and individuals in certain governmental agencies (e.g., individuals and groups in the U. S. Department of Agriculture, state departments of agriculture, mosquito-abatement associations);
8. segments of the media (e.g., chemical-company house organs, chemical journals, farm journals, rural newspapers, radio and TV);
9. elements in some professional societies (e.g., agronomists, entomologists, plant pathologists, weed scientists);
10. a spectrum of private citizens concerned about "threats" to free enterprise and agri-technology and the activities of "irresponsible environmentalists";
11. administrators, elements, and individuals in the land-grant universities, including the Agricultural Extension Service.

I have most likely overlooked additional members of the "club," but it doesn't really matter; the roster is quite impressive as it stands. In many respects, this is an odd set of "bedfellows," but they share a common interest, the pesticide status quo, from which they all think they benefit and which welds them into a powerful political force dedicated to keeping things as they are.

The Pro-pesticide "Mafia"

There is, then, a pro-pesticide "mafia," whose members operate much in the manner of those in its Italian namesake. It has its *famiglie,* its *capi,* its *consiglieri,* its *soldati,* its *avvocati,* its lobbyists, its front organizations, its PR apparatus, and its "hit men." It owns politicians, bureaucrats, researchers, county agents, administrators, and elements of the media, and it can break those who don't conform. In other words, it is a virtual duplicate of the other "*mafie*" that pervade and dominate so much of contemporary American society.

It took me a long time to recognize the existence of the pesticide mafia, and if I had done so earlier in my career I might have been intimidated by it and retreated into my burrow. But now I am too old to care and so I just rear back and blast away at the obscenity. I suppose that this is a dangerous game, but what can a *mafioso* do to an old bombardier beetle except step on it? There are worse fates!

The greed of the pesticide mafia, then, has turned contemporary pest control into a practice in which chemical merchandising has become the name of the game. In fact, the merchandising imperative has assumed such overwhelming influence in our pest-control system that it has made a mockery of scientific pest management. In other words, pest control has become as much or more a matter of moving merchandise as it has of bug killing. As such, it has taken on the major characteristics of the market place: (i) fierce competition between producers of proprietary materials as well as pesticide formulations for a share of the market, (ii) intensive product advertisement by the various companies and the employment of a large sales force to push the merchandise.

As a result of all this, pest control has become a very big business. As best I can determine, over-all insecticide sales in California alone annually approximate $400 million, and application costs probably add another $100 million to the bill. Double these figures to accommodate all pesticides (e.g., herbicides, fungicides, rodenticides) and California's annual chemical control bill adds up to $1 billion, while by my reckoning the national figure totals about $5 billion. Clearly, the pesticide industry has become an enormous one, which in the pattern of our free enterprise economy is compelled to grow. Market stability or regression will not be tolerated in the boardrooms of the American agri-chemical industry, or for that matter, those of Japan, England, Germany, France, Italy, Switzerland, or wherever else pesticides are produced.

Some time ago, a top executive of Chevron Chemical Company made industry's position crystal clear when he told me that unless his firm expanded its markets at a certain annual rate and realized a stipulated profit, the parent corporation (Standard Oil of California) would divert its capital input from pesticide manufacture to other areas of chemical production. Little wonder that under this kind of pressure the pesticide company executive fights to increase his firm's markets and profits. Unfortunately, this market-expansion/profit-making drive, though perhaps commendable in the merchandising of ball point pens, toothpaste, or underarm deodorants, is the worst possible way to go about the business of pest insect management. It is an approach fraught with economic, social, and ecological hazard, and it is a gut issue in the politics of pest control.

It is clear, then, that the agri-chemical industry and its allies have a vested interest in the pest-control status quo (this explains their fierce defense of DDT, which they consider to be the first victim of a conspiracy to banish all pesticides[61]). They have a lot going for them, for they have immense influence over pest-control legislation, pest-control advisement, and pest-control philosophy. Their political muscle is used with great force whenever industry's interests are questioned or challenged. Little wonder, then, that as the dominant stud in the pest-control pasture, the pesticide mafia has compromised or corrupted most of the herd.

The Land-grant Universities

The corruptive and coercive influence of the pesticide mafia is widespread in the land-grant universities, where much of the nation's pest-control research is conducted and from which most of the pest-control recommendations emanate. In the agricultural experiment stations and the Agricultural Extension Service, deans, directors, department chairmen, division heads, or whatever titles they go by, too often knuckle under to the political pressures directly or indirectly generated by the agri-chemical industry and its allies. At their most brazen, those interests have not hesitated to use politically sensitive university administrators to harass fractious researchers. For example, L. D. Newsom, of Louisiana State University, one of America's outstanding entomologists, has been aggressively attacked by four chemical companies in incidents extending over the past twenty years.[62] In each case, industry tried to work its harassment through the highest levels of university administration. The first issue involved Newsom's discovery that one company's insecticide had lost its effectiveness against the cotton boll weevil. Company officials wished to suppress this information and became incensed when Newsom refused to do so.

In the other incidents, including a very recent one, the chemical companies' wrath was incurred when Newsom refused to recommend proprietary products for use on major crops. Fortunately, he is so highly respected in the field and in his university, that the attempts to "get" him have failed. But some of the political bullets, fired with lethal intent, have come close to their mark. Furthermore, even though he has survived, Newsom has had to stand up to virtually continuous badgering for two decades and to commit energy to the time-consuming and mentally wearing defense of his principles.

The second researcher, Denzel Ferguson, formerly of Mississippi State University, was pressured by certain administrators of that institution's College of Agriculture to cease and desist in his opposition to the fire-ant eradication program, and on the same issue was subjected to heavy flak from the Mississippi State Commissioner of Agriculture and from the State Chemist.[63] Ferguson stated in a letter to me that "the President of the University and my immediate supervisors said nothing, because I was tenured and funded with several grants. I would, however, point out that a younger or less well-known person could not have survived the mirex battle. I was simply too well entrenched."

In California, Robert Rudd, author of the highly regarded book *Pesticides and the Living Landscape* did not fare so well. Certain high administrators at the University of California, Davis, objected to his book's message and, following its publication, stripped Rudd of his agricultural-experiment-station title and passed him over for promotion.[64]

Professor Charles Lincoln, of the University of Arkansas, was attacked because he opposed an intensive, season-long cotton pest-control program advocated by a major chemical company.[65] A representative of the company tried to bring pressure against Lincoln through a university vice-president and through a member of the state legislature. Lincoln was also viciously attacked in certain newspapers and farm magazines. Again, as did Dale Newsom, Charles Lincoln survived the ordeal, but one wonders what scars it left.

In a different version of the political pressure game, the Southeastern Branch of the Entomological Society of America was coerced out of promulgating a resolution against the fire-ant eradication program when politicians in Mississippi, reportedly tipped off by a Society member, threatened to cut the Mississippi State University Entomology Department budget and even the entire university budget, were the resolution to be adopted.[66] Not wishing to have a colleague's department and university suffer such punishment, the Southeastern Branch dropped its proposed resolution.

In another incident, when staff members at the University of Arizona initiated and supervised a pesticide-reducing, cost-saving pest-management program in cotton, the state agri-chemical-company organization brought enormous pressure to bear through the highest level of university administration in an attempt to force university withdrawal from the program.[67]

At Texas A&M University, Robert Fleet, a graduate student in the Wildlife and Fisheries Department who opposed the fire-ant eradication program and coauthored an article criticizing it, feels that he lost his research assistantship, was kicked out of his office-laboratory space, and was otherwise hassled and hounded by his superiors, because of his opposition.[68]

What I have just cited is only a sample of the kind of pesticide politics that go on in many, if not most, state agricultural experiment stations; the tip of the iceberg, as the old cliché would have it. What does not show is the implied pressure, even political reprisal, that keeps many, if not most, of the researchers silently toeing the line.

Two incidents will serve to illustrate this point. The first involved a University of California colleague who had become greatly concerned over the heavy spraying schedules forced onto tomato growers by the excessively stringent insect contamination standards set by the food processing industry. This entomologist knew it was impossible to attain the industry-stipulated insect contamination levels in processed tomato products and that, in fact, tiny bits and pieces of insects routinely occur in commercially canned tomato juice, catsup, and spaghetti sauce despite heavy crop spraying. To prove his point he set up an experiment in which he deliberately infested tomatoes with insects, processed and canned them, and then compared the level of insect contamination in his bugged tomato juice with that in canned juice available in the supermarket. He found no difference.

Next, as we university types do in order to inform science and society of our findings and get promoted, he set out to publish the results of his study. But the tomato canners got wind of this and sent a delegation to the university administration to complain about the manuscript and to threaten withdrawal of their grants were the paper to be published. The university brass, upset by this prospect, suggested to the entomologist that he back off. His description of his reaction to this subtle administrative arm-twisting reflects the widespread reality of life in the agricultural experiment stations: "Hell, Van, what could I do? I was just a little guy raising a family and up for promotion. You better believe I tore up that manuscript."

The second incident occurred during the EPA hearings on DDT, and related to the efforts of the Environmental Defense Fund to obtain testimony from aggie-college entomologists for its case against DDT. It began when I received a phone call from Dr. Charles F. Wurster, of the State University of New York at Stonybrook, an EDF heavyweight. I had worked with Wurster in previous DDT hearings (Wisconsin and California) and was scheduled to testify on EDF's behalf in the Washington, D.C., hearings. However, Wurster felt that EDF needed additional research entomologists to support its case, and asked if I knew of several whom he might approach. This was all he asked: Did I know several entomologists who would simply be willing to discuss with him the possibility of testifying?

I told Wurster that I thought there were a few entomologists around who were brave enough to talk to him, and agreed to feel them out on this possibility. So I went to work on the telephone and lined up about a half dozen bug men who expressed their concern over DDT, felt that it should be banned, and indicated a willingness to talk with Charlie about the possibility of testifying in the DDT hearings. Now, these were all old personal acquaintances; good, solid integrated-control types who, in the close circle of long-standing camaraderie and the glow of a bellyful of beer, bourbon, or burgundy, shake their fists and stomp the floor in their resolve to go out and turn the pest-control scene around. When I talked to them on the phone, they were really charged up with a willingness to voice their anti-DDT convictions on behalf of Charlie Wurster and EDF.

But then, evidently, after they had hung up the horn and their adrenalin had dribbled out, they got to thinking "rationally," and by the time Charlie called them they didn't want to have a thing to do with the DDT hearings.

Why? Because, as Wurster later told me, to a man they expressed fears either of administrative reprisal or of threats to existing or proposed research grants.

Believe me, in the aggie colleges many if not most play the game according to the pesticide mafia's rules!

The U. S. Department of Agriculture

For U. S. Department of Agriculture employees, living with political pressure is simply a way of life. These poor people pay that price from the time they join the organization until the day they are fired, resign, retire, or die. Powerful politicans are forever leaning on the federal bureaucracy, and the whole USDA edifice whips and sways under the blasts of congressional heat.

As a result, only too often, leading USDA administrators have been characterized as much by their skills at living with and pleasing key politicians as by their scientific and administrative abilities. This is reflected in the devastating report of a blue-ribbon committee, of the National Academy of Sciences, that investigated the USDA's Agricultural Research Service (now called the Science and Education Administration).[69] A typical statement in this report lambastes "poor research management, including heavy-handed administration, which has both overdirected research and stifled creativity with a welter of bureaucratic impediments." This heavy-handedness is reflected in the reaction of ARS to its fears of political reprisal if researchers speak or write unfavorably of pesticides. ARS answered this problem by appointing an agency censor to blue-pencil suspect prose or rhetoric from manuscripts and speeches.

I am personally aware of two such instances of censorship. The first involved a manuscript entitled "In Defense of Weeds," prepared by my close friend Dr. Lloyd Andres for the report of a pesticide evaluation task force of which I was co-ordinator. In its virgin form Dr. Andres' paper was a beautiful essay, a virtual classic discussing an innovative approach to weed control. But after its rape by ARS it really wasn't worth printing. The second incident involved a speech by Dr. F. A. Lawson, then leader of ARS's biological control pioneer research laboratory at Columbia, Missouri. Lawson, a highly respected research elder statesman, had prepared a strongly critical statement on prevailing pest-control practices, to be read before a major conference in Florida. He submitted his manuscript for review by ARS editors, and then left on vacation. Upon his return, immediately prior to the meeting, he found the "edited" paper on his desk, virtually gutted of meaningful content. I will always remember his speech, which consisted of an enraged muttering about what had happened to his manuscript, followed by a rapid flipping of the papers—twenty or so pages—with the bitter remark that what was left wasn't worth stating. Lawson had recently recovered from a severe heart attack, and I recall vividly my near terror that in his rage and frustration he would have a second seizure. Fortunately, this didn't happen and he is still going strong, but what a terrible moment it was for a dedicated and respected scientist and for his friends and colleagues in the audience.

Censorship of the manuscripts or speeches of responsible, reputable researchers is the ultimate form of scientific debasement. This is the level to which pesticide politics has driven the ARS. And I hasten to point out that the cancer is much more extensive than the few visible tumors, for the very knowledge that censorship exists automatically eliminates controversial discussion from a high percentage of the manuscripts and speeches under preparation.

It is apparent, then, that pest control in the USDA's Agricultural Research Service (Science and Education Administration) is rife with politics, and nowhere is the political evil more manifest than in the pest eradication and area control programs.

The futile fire-ant eradication program, mentioned previously, capsulizes much of this evil. The fire ant, a feisty little beast that invaded the United States from South America some forty to fifty years ago, quickly moved out of its bridgehead and now occurs over virtually the entire Southeast. It is a bothersome animal, having a nasty (though rarely serious) sting, and its nesting mounds speckle agricultural lands, sometimes causing slight yield reductions or damage to equipment when unwary farmers bang into them with their machinery. There are claims, too, that the ant at times attacks and kills nestling birds such as the young of that sacred species of the southern aristocracy the bobwhite quail. However, as a hazard to man, the fire ant is much less dangerous (two recorded deaths up to 1967) than the honeybee or some of the native wasps.[70]

This is essentially the case against the fire ant, a very minor crop pest, an occasional annoyance, and an extremely rare killer of man and wildlife. Adding to the irony of the situation, the fire ant's bad traits are in fact substantially balanced by the good it does through predation on cotton pests.

But southern folks don't like the fire ant, and out of this pique southern politicians, USDA bureaucrats, and involved chemical-industry personnel fashioned a costly chemical control program and a major ecological threat. Many entomologists realized this, and so did conservationists and environmentalists. In fact, in 1967 concern over the fire-ant eradication program led to the appointment of an ad hoc committee of the National Academy of Sciences to investigate the program's feasibility. This committee, composed of an elite group of America's entomologists as well as other leading scientists, went about its assignment in an energetic and highly competent manner. Its thorough and penetrating report recommended against the program, stating that the fire ant was not an important pest, that the eradication effort was unlikely to succeed, and that limited local control measures would be adequate.[70] However, the USDA ignored these recommendations, a fact that was not generally known until years after the report was submitted. And so the program went on. But eventually the heat generated by increasing public concern over the widespread dumping of the "eradicant" chemical mirex into the environment, and its failure to eliminate the ant, occasioned a reassessment of the program. Unfortunately, this led only to minor changes, such as the alteration of the program objective from eradication to control, and some restrictions as to where the mirex granules were to be scattered. Thus, for years on end, to the tune of approximately $10 million per year, the public supported a pollutive area-control campaign against a minor nuisance, while the involved politicians, bureaucrats, and chemical-company officials chortled over their slick deal, which to date has cost the public more than $150 million.

Interestingly, in 1975 the USDA, complaining that it could not properly attack the fire ant under prevailing restrictions imposed by the Environmental Protection Agency, suspended its control program. USDA was, of course, only playing games. It fully intended to get back into the fire-ant act and was simply waiting for political pressure to force EPA to back off. But its plans were rudely sidetracked by the Kepone® scandal. It just so happens that mirex is a sibling—a nearly identical twin—of Kepone®, and the political and legal heat generated by the Kepone® tragedy in Virginia prompted the Allied Chemical Company to drop mirex production, and it also forced the USDA Animal and Plant Health Inspection Service (APHIS) entomologists, who mastermind the fire-ant program, to duck for cover. However, with pork-barreling southern politicians itching to get their $10-million-a-year welfare program back in gear, and barring a permanent ban on mirex, we can expect APHIS to surface again with the mirex miracle once the Kepone® horror has faded from memory.

The influence exercised by politicians, industry, and bureaucrats over federal pest-control spending was dramatically brought home to me several years ago, when I was on a task force advisory to the President's Council on Environmental Quality. A statistic that surfaced during one of our deliberations shocked me at that time and has remained in my mind ever since. This related to the annual government (USDA) expenditures on pest "eradication" and "area control" programs, versus those for all entomological research. As I recall, the figures for the particular year were approximately $31 million for eradication and control, and $19 million for research. In other words, politically inspired, largely futile, frequently pollutive eradication and control programs that, as the Office of Management and Budget was grumbling, never seem to terminate were receiving two thirds again as much support as pest-control research, whose goal is problem solving.

There is no reason to believe that the formula has changed—and in fact, with the boll weevil eradication program (projected at full cost to amount to something between $0.6 billion and $1.6 billion) still a possibility and with increasing pressures generating out of the screwworm, gypsy-moth, tussock-moth, and spruce-budworm problems, matters may well worsen. In other words, at the federal funding level, pest-control pork-barreling is rampant. The same is true in many of the states. For example, in California alone the State Department of Food and Agriculture is expending millions of dollars annually on pest eradication and containment programs. One of these, the grape-leaf skeletonizer eradication program, which had been going on since the early 1940s, cost California's taxpayers $660,000 in 1974, before Governor Brown cut it out of the budget.[71]

Pest eradication and containment is largely welfare in the guise of pest control, and as such it is the essence of the politics of pest control.

The Professional Societies

Pest-control politics has not only corrupted governmental agencies and educational institutions, but its toxins have also permeated the professional societies. Take the Entomological Society of America (ESA) for example. Ever since the old Entomological Society sacrificed much of its dignity and gave its name to a coalition with the American Association of Economic Entomologists, things have been in a bad way. The low point was reached in 1964, when the ESA passed a resolution condemning the journal *BioScience* and ecologist Frank Egler for an Egler-authored article entitled "Pesticides in our Ecosystem".[72,73] This event was so disgusting that it goaded famed ecologist Lamont Cole of Cornell University into one of the most eloquent put-downs I have ever read.[74] Cole roared, "In my twenty years of association with scientific editorial boards and publication committees, including a five-year term as a senior editor, I have been on both the sending and receiving ends of letters criticizing the acceptance of particular manuscripts. But it is something entirely new in my experience for a scientific society to pass a resolution condemning the editors of a scientific journal for granting a recognized senior scientist the right to express his views in print. This extraordinary event and the opposing forces involved call for scrutiny by the scientific community."

Cole went on to ask whether the Entomological Society is dominated by economic entomologists. Here he was a bit off the mark, for he had, in fact, identified the culprits earlier on in his letter as a coalition of "chemists, toxicologists, and others primarily concerned with the destruction of insects." There is quite a difference between these people and economic entomologists. Many of the latter, like the harassed Dale Newsom and Charles Lincoln, are outstanding insect ecologists and developers of rational pest-control programs.

The influence of the pesticide proponents over the ESA is further reflected in the list of invited speakers to the national meetings of the Society. In this connection, it is interesting to note that since the 1962 publication of Rachel Carson's *Silent Spring,* not one of the strong critics of the pest-control status quo has spoken before a plenary session of the Entomological Society or participated in one of its major symposia. In other words, members of the Society have never heard any of such prime movers of pesticide reform as Barry Commoner, Rachel Carson, Robert Rudd, Frank Graham, Victor Yannacone, Charles Wurster, Robert Risebrough, Paul Ehrlich, and Ralph Nader. Instead, they have been treated to the sage observations of such outstanding advocates of rational pesticide usage as Lea Hitchner, President of the National Agricultural Chemicals Association; Congressman Jamie Whitten, good friend of the agri-chemical industry and author of the volume *That They Shall Live,* the industry-blessed "rebuttal" to *Silent Spring;* James T. Conner, chief Washington lobbyist for the National Agricultural Chemicals Association; and a series of representatives of the Plant Pest Control Division (now called APHIS), the USDA's scorched-earth insect-eradication agency. The pro-pesticide posture of the Entomological Society has, if anything, become more rigid in recent times. For example, at the Society's 1975 meeting, in New Orleans, two thousand members in attendance at the annual awards luncheon, paid for out of their individual registration fees, were treated to a political tirade against the pesticide-regulation policies of the Environmental Protection Agency by Assistant Agriculture Secretary Robert F. Long, staunch friend of the agri-chemical industry.[75]

Pesticide politics in the Entomological Society of America? You better believe it! And as I will mention later, the same is true in other agri-science societies.

Chapter 7

STICKING IT TO CESAR—THE SOCIOLOGY OF PEST CONTROL

Some years ago on one of those indescribably lovely spring days in the California desert, my entomologist crony Vernon Stern and I were cruising the dusty back roads of the Palo Verde Valley in search of an alfalfa field in which to conduct an experiment. As we drove along a county road a few miles north of the city of Blythe, we came upon a cantaloupe field where a bare-chested man was loading pesticide into the hopper of a parked crop-dusting rig. Since we had been driving for some time without finding a suitable candidate experimental plot, we decided to stop and ask the rig driver whether he knew of some nearby alfalfa fields.

The man, a *bracero,* or "wetback," with work to do and intent on getting it done, nevertheless smiled as we approached and halted his labors as we asked in a mixture of English and Spanish whether there were any alfalfa fields in the area. He didn't know, since he was new to the ranch, having just replaced another worker, who had fallen ill. We thanked him, and then just to make conversation, asked him what pest he was dusting.

"*Pulgones.*"

"Oh, aphids. What pesticide are you using?"

"I don't know, I can't read the label; just some medicine for *pulgones.*"

As is the habit of our breed, Stern and I automatically flicked our eyes to the label on the pesticide sack. It read, PARATHION. Parathion is one of the deadliest nerve-gas derivatives among the modern insecticides. We were stunned. Here was this smiling, bare-chested laborer, his body frosted with parathion dust, breathing it in and licking it off his sweat-moistened lips, totally ignorant of his peril. Little wonder his predecessor had fallen ill! As best we could, Stern and I implored him to immediately stop his dusting activities, strip off his remaining clothing and jump into the nearest irrigation ditch to wash off the poison. But his response was a friendly laugh, an *adiós,* and the resumption of his crop-dusting activity. He was a happy young man, with a well-paying job, a boss to satisfy, and no more time to waste with a couple of silly *gringos* all worked up over some bug medicine.

When we got back to town we reported the matter to the local agricultural authorities, who, I am quite sure, never did a thing about it. But even today, years later, I occasionally fret over that cheerful Mexican youth and wonder how long it was before he, too, became ill and gave way to an equally innocent successor.

This anecdote illustrates an ugly facet of the sociology of pest control. Too often the growers' thought seems to be, to hell with the hazards; just kill the damned bugs and get on with producing the crop. United Farm Workers President Cesar Chavez has long had a different viewpoint on this matter, and through his union has attempted to bring some kind of order out of the chemical chaos. The attempts have come via litigation, legislative action, and stipulations in union contracts, and of course they have met stiff opposition.

Chavez told me of his concern about pesticides one day

when I had come down to UFW headquarters in Delano, California, to act as a resource person on the union's behalf during a visit by Senator Walter F. Mondale, then chairman of the U. S. Senate Subcommittee on Migratory Labor.[76] Another pesticide specialist was on hand, and along with Mondale and his aide, we interviewed UFW personnel including Chavez (then bedridden with a back ailment) and visited union facilities. As part of their effort to impress upon Mondale the seriousness of the pesticide problem, Chavez and his UFW colleagues had arranged to have on hand a number of farm workers who had directly suffered pesticide-caused injuries. This group, as would be expected, was overwhelmingly Chicano, but much to my surprise, one was an Anglo, a grizzled, slope-shouldered old Okie, who told the saddest story of all.

At the time of the Mondale visit, this man was totally work-incapacitated by a respiratory ailment, which he felt had been severely aggravated by his having been required to work with hazardous pesticides. In this connection, he was particularly bitter about an incident in which he had been forced by his rancher-employer to continue spraying a vineyard with a dangerous insecticide even after he had complained to the rancher that the chemical made him ill. In taking his complaint to the grower, the worker had been perceptive enough to associate his illness with the insecticide he had been using, to check the label, and to determine from appropriate sources that the material was indeed hazardous. In light of his illness and having satisfied himself that the insecticide was dangerous and probably the reason for his not feeling well, he asked his boss to transfer him off the spraying assignment. The grower responded by telling him to report to the spray rig the next morning or get off the ranch. The deep irony of this tale is that the man had been working on the ranch for seventeen years!

Such is the sociology of pest control!

Chavez and his UFW colleagues became increasingly concerned over pesticides in the late 1960s as greater knowledge unfolded concerning the hazards they posed. Eventually the union made an effort to obtain insight into the kinds of pesticides being used, the amounts applied, and the places and times of application. The initial effort to gain this information was through the office of the Kern County, California, agricultural commissioner, the keeper of official pesticide use records in that major crop-producing county. The effort was totally frustrated by one of the most shocking acts of collusion between public officials and a vested interest of which I am aware.

I learned of this collusion in a most interesting way, from a Commissioner's Office staffer, in the Kern County Superior Court, in Bakersfield, in January 1969. But first I should explain how I got there.

This came about apparently because of my growing reputation as a critic of the pesticide status quo. At any rate, one day in late 1968 David Averbuck, a lawyer with the United Farm Workers Organizing Committee (UFWOC), showed up at my office in the Division of Biological Control, University of California, Berkeley, and told me of the union's concern over the pesticide hazard to farm workers, its attempts to gain access to the Kern County records, the agricultural commissioner's refusal to produce the records, and the impending court hearing on the matter in Bakersfield. He asked if I would testify on UFWOC's behalf. I agreed, because I was aware of the hazardous nature of many of the insecticides and the sloppy way in which they were being used. I was especially concerned because some of the assistants who had been working in our university experiments, where we used the organophosphate methyl parathion, had suffered severe depression of blood choline esterase despite the careful safety measures we employed. Choline esterase is an enzyme involved in nerve message transmission. Many of the modern insecticides are choline-esterase inhibitors and hence kill by disrupting nerve transmission. It may be a blow to our ego, but these insecticides work on us in exactly the same way they do on the insects, and that is why they are so dangerous. My concern was that since methyl parathion had affected our carefully supervised assistants, things were almost certainly far worse with the farm workers, operating under conditions that were much more hazardous. Here the memory of the parathion-dusted Mexican lad in the Palo Verde Valley came vividly to mind.

On the morning that I stepped into the Bakersfield courtroom, I was confronted by a scene in which to my right the seats were occupied by persons obviously associated with or sympathetic to the farm workers; that is, Chicanos, Filipinos, and young, militant, hippie-ish Anglos, while to my left the group was mainly composed of prosperous-looking, conservatively dressed, neatly groomed, WASP-ish males, among whom I quickly recognized a number of agribusiness types and county, state, and university employees.

As I was surveying the scene, one of the people on the left waved and beckoned me to sit next to him. I immediately recognized him as a high Kern County agricultural official whom I had known for years, and unhesitatingly joined him. I had hardly settled into my seat when he nudged me and pointed to a gentleman on the other side of the aisle and said, "See that son of a bitch, that's Jerry Cohen, lawyer for UFWOC. He came into our office a while back and wanted to look over our pesticide use records. We refused to let him do it. When we did this, he took off, saying he was going to seek a legal order to force us to open our files. Well, we fixed that. We got on the phone and called the ag-chemical people and asked them to sue us, to keep our files closed."

This is apparently what had happened: the Agricultural Commission's Office contacted an agri-chemical representative, told him what Cohen and UFWOC were up to, and suggested that one or another of the agri-business groups take legal action to enjoin the agricultural commissioner from opening his files to UFWOC. Shortly thereafter, the Bakersfield Superior Court held a hearing concerning *Atwood Aviation Inc.* (crop duster) *et al.* v. *Seldon C. Morley* (agricultural commissioner), with Cohen and Averbuck as intervenors.

To complete the story, the court's decision went against UFWOC. That is, Atwood Aviation et al. were upheld by the court in their effort to prevent UFWOC from gaining access to the agricultural commissioner's files. Reason: possible disclosure of trade secrets!

Later I participated in a similar episode in the Riverside County Superior Court, and again the Farm Workers lost. But this decision was appealed and ultimately overturned by the Fourth District Court of Appeals, Division Two. This means that farm workers do indeed have the right to know what kinds of poisons are being applied to the fields and orchards in which they work.

Some feeble advances have occurred since these two trials and the reversal of the Riverside Superior Court decision. The California Department of Food and Agriculture now has a computerized pesticide-reporting system, which gives a quarterly crop-by-crop summary of pesticide use and (one hopes) may someday provide a field-by-field breakdown. Re-entry protocols have also been set for several pesticides in some crops. That is to say, there is now the beginning of a system that assures that no human beings, not even the heretofore expendable Chicanos, can be sent into poison-doused fields or orchards until sufficient time has elapsed for the toxic residues to dissipate. Upgrading of pest-control advisers and increasing implementation of integrated control systems may lead to reduced and more civilized insecticide usage. If that doesn't work, perhaps the increasing cost of insecticides and shortages of material will help. The United

Farm Workers have insisted on safe pesticide-use clauses in their contracts. As the union grows in vigor, this should lead to even safer practices.

But, returning to the Kent County trial, two things remain indelibly etched in my mind. The first is a memory of the fear and hatred that the dominant San Joaquin Valley middle-class establishment holds for Cesar Chavez and his United Farm Workers, and the impression that this middle class considers the Chicano, Okie, and black rural population to be somewhat outside the pale of humanity. The second memory is of a corrupt act in which public officials colluded with one element of the citizenry against the rights and well-being of a less-advantaged group.

The sociology of pest control is indeed an ugly game.

Chapter 8

THE TERRIBLE TUSSOCK TUSSLE

In the battle over pesticide regulation, the pesticide mafia took its stand with DDT. The reasoning was quite simple: if DDT were to be shot down, an array of pesticides would fall in its wake. Norman Borlaug, a vociferous DDT supporter, spelled this out in his famous 1971 FAO speech in Rome when he likened DDT to the first of a series of tumbling dominoes.[77]

The DDT showdown occurred in the hearing rooms of the Environmental Protection Agency in Washington, D.C., in 1971–72. The pesticide's supporters were well prepared for this battle, marshaling their shrewdest strategists and heaviest weapons (including Borlaug). They had a lot going for them and seemed determined to break the backs of the eco-freaks once and for all and get on with bug killing. But EPA and its redoubtable ally the Environmental Defense Fund were even better prepared and won the day, which culminated in William Ruckelshaus' courageous decision to ban DDT (Ruckelshaus had guts even before the Saturday-night massacre).

This was a crushing setback for the pesticide mafia, but as so often happens with a battered force, considerable sting remained in this chemical scorpion. In fact, it responded to the Ruckelshaus decision with the most wicked bolt it had yet unleashed, the tussock-moth ploy.

The Douglas-fir tussock moth is a native insect that periodically occurs in outbreak numbers in western forests. Its larvae feed on the needles of valuable Douglas firs and true firs, sometimes stripping and killing the trees. Foresters hate it, forest products companies hate it, small timber owners hate it, logging-industry workers hate it. All have a consuming passion to kill it. This passion rose to unprecedented intensity in 1972, '73, and '74, when an extensive outbreak of tussock moth occurred in the Pacific Northwest. And coming as it did in DDT's gravest hour, this event was a godsend to the proponents of that insecticide. Quite predictably, they jumped at the opportunity to promote DDT as a tussock-moth panacea and made it the bone of contention in the terrible tussock tussle. This was the tussock-moth ploy.

As a last-ditch DDT support weapon, tussock moth was a sinister missile with the capacity to create enormous mischief. Most importantly, it kept DDT's foot in the door and gave its proponents time to hold the line, regroup, and pump new life into their cherished biocide.

The tussock-moth ploy took advantage of a provision in the Ruckelshaus decision that permits use of DDT in the event of a public health crisis or impending economic disaster where there is no other effective deterrent. What made tussock moth so important is that unlike other loophole cases (e.g., control of mice, bats, pea-leaf weevils), it was a big deal that could be exploited politically and in the media. As mentioned, the tussock moth can strip fir trees of their needles, leaving them skeletal over thousands of acres. When it occurs, this is highly visible damage to which the spray advocates point with alarm while loudly decrying the loss of valuable timber and the desecration of magnificent forests. Never mind that this has been going on for millennia, that the problem has been aggravated by "high grade" logging (which produces susceptible stands), that much of the defoliation occurs among trees on poor growing sites, and that most trees withstand the attack and grow with renewed vigor.[78]

What mattered with tussock moth was that a mixed bag of individuals and agencies with vested or emotional interests in DDT had found a perfect vehicle to promote their cause. All they had to do was convince enough people that DDT was the only way to go in tussock-moth control. In other words, tussock moth provided a golden opportunity to pull the DDT skunk out of the EPA garbage can, and its supporters wasted no time in rallying various groups to effect the rescue. This bizarre lineup included the forest products (logging) industry, which wasn't about to risk a single log in the interest of ecology when DDT, with the government footing much of the bill, just might clean up the bugs; small-time timber owners, who feared that their individual stands might fall in the pathway of the dreaded insect horde; loggers and other logging industry workers, who were gulled into believing that the tussock moth was about to gobble up their jobs; Forest Service administrators, reflecting the parent U. S. Department of Agriculture's pique over the DDT ban (USDA, jabbed in the derrière by farm state politicians, had supported DDT in the EPA hearings) and their own resentment of "those eco-freaks" dictating what could or could not be sprayed onto the forest; local politicians (including congressmen), beholden to the forest products industry giants; powerful southern congressmen, traditional darlings of the agri-chemical industry, who were eager to exploit any opportunity to get DDT back on track; and the regional press, unquestionably reacting to the desires and muscle of its potent client, the forest products industry.

The poor forest never had a chance, nor did its concerned allies: EPA, the conservationists, and the overwhelming majority of forest entomology researchers. Nevertheless, EPA, acting in the public interest, made a noble effort to uphold the DDT ban, but it inevitably knuckled under to the crudest sort of blackmail, generated by the politicians.

The hatchetman in this disturbing episode was Congressman W. R. Poage (D-Texas), chairman (since deposed) of the House Agriculture Committee, who announced that if EPA failed to approve a Forest Service request for the use of DDT against the tussock moth, the committee would immediately seek House action on a bill stripping EPA of its authority to regulate use of the compound. Even Gerald Ford, then Vice-President, added his voice to this power play by stating, "If they (EPA) don't respond to a problem of this sort, I think Congress might change the law."[79]

This was forceful arm twisting and it brought EPA to heel. The implication was crystal clear: "Shape up, EPA, or you not only lose DDT regulation but the whole pesticide registration and regulation bundle as well!" Stripping EPA of its pesticide watchdogging role is a major goal of the pesticide promoters and their politician lackeys, who would dearly love to return this "responsibility" to the compliant hands of the U. S. Department of Agriculture. In fact, Congressman Poage subsequently sponsored legislation that was designed to give the USDA virtual veto power over EPA pesticide regulation decisions (see Chapter 11).

EPA Director Russell Train had no choice but to capitulate, for if he did not, the agency would have faced virtual castration. EPA had already lost its bid to regulate atomic energy, had been overridden in its efforts to establish meaningful automobile emission standards, and was being challenged on its stand against offshore oil drilling. Now, if its pesticide regulating authority were removed, it might as well close shop. So Train apparently struck a bargain with the DDT muscle men. In exchange for the continued right to regulate DDT, EPA would permit its use against the tussock moth.

I first learned what was up months before the proposed public hearings on the use of DDT. This information came from associates in the Environmental Defense Fund who told me that they were backing off from the tussock-moth issue because the politicians had made it clear that if EDF rescued the situation, EPA would be held to account. The DDT boys had placed their roadblocks very cleverly! EPA then proceeded with the tussock-moth hearings as though they really mattered. This exercise was ostensibly conducted to air both sides of the issue and thereby help Train "make up his mind." It also gave an aura of credibility to what was coming.

I recall mentioning the farcical nature of the hearings to an EPA official who had phoned from Washington, D.C., asking me to testify in support of the DDT ban.

He acknowledged that there was enormous political pressure to unfetter DDT and that things looked grim, but he insisted that the cause was not lost. I felt at the time, as I do now, that he was acting out his part in the farce. Whatever the case, EPA was seeking window-dressing support for the DDT ban because the Forest Service, playing it super safe, would not allow its own researchers to testify. I convinced the EPA official that, as a non-forest entomologist, I might be discredited, and suggested, instead, that he approach tussock-moth researchers Donald Dahlsten, a Berkeley colleague, and Steven Herman, of Washington's Evergreen State College. He accepted my demurrer and contacted Dahlsten and Herman, both of whom testified in a hearing held at Portland, Oregon. The two were aware of the hopelessness of the situation, even as they agreed to testify, but they went through with their act as a matter of principle and because they were confident that events in the field would prove them right. In this latter light, it was important for them to have their opinions on record regardless of EPA's forced decision. It is always satisfying in such matters to resurface and confront the opposition with fully documented "I told you so's."

The most disturbing aspect of the entire affair was the Forest Service's muzzling of its research entomologists. An overwhelming majority of these researchers opposed the use of DDT. They were concerned about the material's ecological impact and especially upset because they knew that the tussock-moth population was already collapsing of natural causes and didn't require wide-scale spraying. What's more, many were confident that alternative control materials were available for the limited spraying that might be necessary. These people also knew that the threat posed by the pest had been grossly overblown, that much of the "severely threatened" fir forest was on poor growing sites, and that the bulk of the damaged trees would probably refoliate ("green up.")

But the Forest Service administration had the political backing for its DDT stand and saw to it that its dissenting researchers did not testify at the several EPA hearings.

And so, after all of the huff and chuff of the hearings, an irascible Russell Train issued his coerced order permitting use of DDT against the Douglas-fir tussock moth. It was in making this statement that he charged the Forest Service with virtual dereliction in failing to develop alternatives to DDT.

As matters turned out, the spray program was a fiasco. Dahlsten and Herman and the muzzled Forest Service researchers were vindicated when the tussock-moth population suffered a natural collapse even as the spray planes dumped their unneeded pollutant. Now we look at the cost of this bit of political chicanery:

1. At least $3 million expended in spraying a hazardous chemical on 427,000 acres of forest and its contained plant and animal communities to "control" a phantom pest population;[80]

2. Numerous non-target creatures including untold thousands of birds destroyed by the biocide;[81]

3. Eighteen thousand cattle and several hundred sheep rendered unmarketable by DDT contamination;[82]

4. Game species so heavily contaminated with DDT that hunters had to be warned against consuming the meat of animals they might bag;[83]

5. Expenditure of additional government funds to compensate the Coleville Indians while their DDT-contaminated cattle were held off the market;[84]

6. Neglect of research on DDT alternatives, including two chemical insecticides and two microbial materials, while the Forest Service poured its millions of dollars into the DDT spray program;[85]

7. The emboldening of pest controllers to consider DDT's use in other forest pest problems;[86]

8. Maneuvering by the pesticide mafia to increase agricultural use of DDT.[87]

The full irony of the situation is summed up in the words of Dr. Robert E. Buckman, Director of the Pacific Northwest Forest and Range Experiment Station. Buckman, a bureaucrat going about as far as he dared, told the Western Forestry and Conservation Association's 1974 convention, in Spokane, that the tussock-moth population had already been suffering natural collapse when the Forest Service conducted its massive DDT spray program. He glossed over this shocking admission by stating that several chemical and biological alternatives to DDT showed promise and predicting that during the next tussock-moth outbreak DDT would probably be supplanted by a more desirable alternative.

But the question asked all along by the conservationists and forest entomology researchers, "Why was the spraying undertaken at all?" went unanswered. There was simply too much heat on the Forest Service to permit such candor.

Even in its experimental evaluation of alternative insecticides, the Forest Service could not break away from its pro-DDT prejudice. The experiment was planned in such a way that DDT would almost surely outperform the other materials, particularly carbaryl and trichlorfon. Dr. Carroll Williams, the Forest Service research entomologist who did the testing, warned of flaws in the experimental design before the study was undertaken. Williams' warning was ignored, as was his disclosure that tussock-moth populations were crashing in the proposed experimental area, and his suggestion that the study be moved to an area of viable population if it were to have meaningful results. Since DDT was applied in higher volumes and with better coverage than were carbaryl and trichlorfon, of course it outperformed these two materials. Nevertheless, they still killed a substantial percentage of the larvae. In fact, trichlorfon gave foliage protection equal to that provided by DDT.[88]

The greatest mystery of the tussock-moth episode concerns just how much timber was actually destroyed. In this connection, there are some interesting statistics to show the degree to which the problem was overblown. At the height of the tussock-moth alarm, the Forest Service claimed that about eight hundred thousand acres of prime fir forest was heavily infested and that if this acreage was not sprayed with DDT there would be extensive tree mortality. But while these disaster warnings were being sounded, the research entomologists who best knew the situation maintained that the seriously threatened acreage amounted to only a fraction of the "official" estimate. They also felt that this limited high-hazard acreage could be identified and selectively sprayed. In this latter connection, a number of researchers felt that the microbial insecticide *Bacillus thuringiensis* and the chemicals carbaryl and trichlorfon would give adequate control.[89]

The researchers' estimate of the problem seems to have been quite accurate. This is supported by the remarks of Dr. David Graham, the Forest Service's co-ordinator of the DDT spray program, who in the July 16, 1974, issue of the Portland *Oregonian* is quoted as saying that eighty-eight thousand of the originally threatened hundreds of thousands of acres had been destroyed by the tussock moth.

Dr. Graham was a vigorous proponent of the spray program, and therefore it must be assumed that his figure for tree loss is generous. Yet the eighty-eight thousand acres represents only about 10 per cent of the originally estimated "gravely threatened" area. The point here is that if the Forest Service had been given a free hand, more than three quarters of a million acres of forest would probably have been sprayed with DDT to protect the eighty to one hundred thousand acres that actually contained seriously threatened trees. The delaying tactics of EPA averted this gross environmental insult, but only in part, as the sprayers finally had their way on 427,000 acres in 1974. As regards timber loss, it is interesting to note that no one really knows what it was, because the Forest Service began jerking trees out of the forest in its "salvage" program almost before the tussock-moth larvae had stopped munching foliage. Reportedly, many of these "salvaged" trees would have "greened up" if they had been left standing, but what is even more distressing is that many others were perfectly sound, having suffered little or no damage. In other words, there is reason to believe that the Forest Service fattened its tussock-moth loss statistics by chopping down healthy trees under the guise of salvage.

There is a final facet to the tussock-moth episode, which in many respects is the most distressing of all. Here I have in mind the failure of the Forest Service to anticipate the

1972 outbreak. This is not surprising, because the Service's Pest Control Division had never bothered to devise a tussock-moth early warning system. Indeed, such technology is probably beyond its capability, and so Smokey Bear was asleep at the switch when the insect began to crank up its larval legions. By the time the Service, with its inadequate detection program, realized that a crisis was at hand, the forest had already begun to frazzle under the lepidopterous assault, and most of the trees that were to die during the outbreak had already been fatally injured.[90] Thus, even if DDT had been instantly available and effective, it would not have saved the situation. This makes all the subsequent bombast, hand wringing, and political chicanery generated by the DDT issue a completely hollow exercise.

The two-little and too-late pattern has characterized Forest Service action in every tussock-moth outbreak on record. But in 1972 the problem was too big to be swept under the carpet, and so a scapegoat had to be found and quickly.

Ergo, EPA, the DDT foot-dragger.

This all points up the fundamental weakness of the pest-management strategy of federal, state, and most private pest-control practitioners. These persons, though excellent bug killers, skilled in the logistical and toxicological aspects of their calling, lack ecological sophistication. This is acknowledged by forest entomologist Graham and his colleague K. H. Wright, who told the Northwest Forest Pest Action Council, ". . . We do not have an adequate system for measuring, evaluating, or predicting insect- and disease-caused impacts. . . . Basically we do not have a clear understanding of the concepts and practical implication of pest impacts in the total space-time frame of the resource management process." In short, the bug killers simply do not understand population ecology and ecosystem dynamics, and lack the ability to conceptualize and implement problem-solving programs. Instead, they employ one or the other of two tactics in a simplistic chemical control strategy. The first is area-wide preventive treatment; the second, fire-brigade spraying in response to bug outbreak alarms, the tactic employed against the tussock moth. But regardless of the tactic, the game never ends. And why should it? It is the raison d'être of the pest-control practitioners. In the private sector we see it played in the massive spraying of our agricultural crops; in the governmental area, in programs against such pests as spruce budworm, fire ant, gypsy moth, boll weevil, and, of course, tussock moth. The tussock-moth ploy was a tragic hoax masterminded by the friends of DDT. Predictably it backfired, and we now have the opportunity to profit from the debacle.

First we can insist that it never again happen with tussock moth, but more importantly we should take a very hard look at the pollutive, resource-wasting, energy-gobbling, chemical pest-control strategy and come up with something better. But this will not be easy. The pesticide interests continue to have enormous influence over American and global pest-control policy, and one of their highest priorities is to get DDT back into the mainstream. There can be little doubt that they operated behind the scenes in efforts to revive DDT use in Louisiana cotton and that they will continue to precipitate and support similar maneuvers in the future. In this connection, it will be very interesting to watch developments in the spruce-budworm situation, where the first faint signals of a revived DDT campaign can already be detected.

DDT symbolizes the pest-control status quo, and because of this it will continue to be the focus of savage conflict between those who are seeking change and those who want things to remain as they are. The evolution of a rational pest-control strategy very much depends upon the outcome of this conflict.

Chapter 9

THE INSTANT PROFESSIONALS*

Perhaps the greatest absurdity in contemporary pest control is the dominant role of the pesticide salesman, who simultaneously acts as diagnostician, therapist, nostrum prescriber, and pill peddler. It is difficult to imagine a situation in which society entrusts so great a responsibility to such poorly qualified persons. Pesticides rank with the most dangerous and ecologically disruptive materials known to science, yet under the prevailing system these biocides are scattered like dust in the environment by persons often utterly unqualified to prescribe and supervise their use.

Pest-control advisement should be a high-grade technology conducted by thoroughly qualified technicians. Instead it is overwhelmingly in the hands of skilled merchandise hucksters employed by the agri-chemical industry. Little wonder that contemporary pest control is characterized by economic, ecological, and social chaos.

It isn't as though thought hasn't been given to the matter and attempts made to turn things around. The simple truth is that the pesticide mafia knows that its salesmen assure its domination of pest control, and it keeps a militant watch on

* Excerpted in large measure from an article published in the April 1973 issue of *Organic Gardening and Farming* magazine.

this aspect of the status quo. Thus, it was right on the scene in California a few years ago when attempts were made to develop a meaningful pest-control adviser's examination and licensing law. The pesticide mafia sized up the situation, recognized that a good law would be a threat to its interests, and molded its own straw-man law to serve its purposes. Let me tell you how this was done.

During the late 1960s, in an atmosphere of ecological concern, heavy pressures developed in California for the enactment of legislation to require the examination and licensing of pest-control advisers. As a result, in 1970 State Senator (now Congressman) Anthony Beilenson proposed an excellent piece of legislation that in addition to its examining/licensing requirement would have (i) prevented persons affiliated with pest-control companies (salesmen) from recommending the use of legally defined injurious materials (most modern insecticides), and (ii) excluded chemical company employees from eligibility to serve on the state pest-control advisory committee. With these provisions in it, the bill, if enacted, would have been a giant stride toward the evolution of a rational pest-management system, for it would have broken the dominance of pesticide merchandising over pest-control practice.

Senator Beilenson, who was considered by many to be the outstanding member of the California Legislature, lived up to that standard in his approach to the pest-control-advisers bill. Thus he did not act covertly, but instead, through his legislative assistant, Robert Toigo, openly contacted University of California pest-control specialists for technical advice. The group contacted was no gaggle of eco-radicals, but, as I recall, included Dr. Ray F. Smith, Chairman of the Berkeley Entomological Sciences Department; Dr. William W. Allen, a Berkeley research entomologist of conservative leanings; Dr. Clarence S. Davis, an extension entomologist; Dr. Louis A. Falcon, an insect pathologist; and myself.

The Beilenson bill very nearly attained enactment, but then, at a late hour, the pesticide mafia, sensing the threat that the bill posed to its interests, effected its abortion. Furthermore, the shrewd *capi*, realizing that it was only a matter of time until someone else's examining/licensing law was passed, moved quickly to fashion a bill of its own. Ergo, the straw man.

In executing this rip-off the pesticide mafia moved swiftly and furtively to get official blessing for its bill. It apparently did this by proposing to friendly top-level administrators within the University of California and the California Department of Food and Agriculture that they appoint a joint committee of specialists to draft a background document for the proposed bill. This the administrators dutifully did, with the committee being composed of a university researcher, an extension specialist, and a county agricultural commissioner. Once the background document was completed, it was passed on to virtually every group interested in maintaining the pesticide status quo, for them to take their shots at it. As could be expected, the riddled version of the original background document gave the mafia and its politician allies all the basis they needed to shape a law that makes a mockery of the examination/licensing intent.

Under this law, pesticide salesmen are included without restriction among the licensees, and the Pest Control Advisory Committee (read board of examiners) includes a representative of the pesticide industry, a licensed pest-control operator (i.e., crop duster), and a licensed agricultural pest-control adviser (most likely a salesman, since salesmen account for more than fourteen hundred of the approximately eighteen hundred fifty licensed advisers[91]). This is equivalent to a board of medical examiners having a drug salesman, a bedpan manufacturer, and an ambulance driver among its members. The examinations for the various specialties listed under the law are incredibly simple. For example, my old friend and former university colleague Everett Dietrick, who now conducts an insectary business, told me that his secretary, acting on a dare, boned up for one of the examinations, passed it, and is now one of California's licensed pest-control advisers. Barry Wilk, a Berkeley entomology staff research associate and graduate student, told me that he also passed one of the examinations, which he termed a joke. He also said that a farmer friend who had formerly been a pesticide salesman offered him a stack of old examinations just in case he wanted to get a preview of what was coming. This is most interesting, because officials in the California State Department of Food and Agriculture, pleading their need for secrecy, refused to make copies of old examinations available to me and a team of coinvestigators when we were studying California pest-control practices under an EPA contract.

The examination and licensing law has been a severe setback to the development of a rational pest-control system in California, because it drapes the pesticide salesman with a mantle of professional respectability and thereby enhances the myth that he offers competent and objective advice on pest-control problems. Now when the salesman flashes his business card to a prospective customer, it bears the impressive title LICENSED PEST CONTROL ADVISER, and this title is backed by a document bearing the seal of the great state of California. The salesmen are so proud of their newly achieved respectability that they have formed an organization, the Council of California Agricultural Pest Control Advisers, to advertise their transition from peddlers to "professionals." But despite their instant professionalization, they remain salesmen, and rational pest control suffers because of their legally sanctioned camouflage.

California's Agricultural Pest Control Advisory Committee, with its inclusion of chemical company employees and a pest-control operator (spray applicator), fortifies the misconception that pest control and chemical control are essentially synonymous. What other conclusion can be reached so long as the Pest Control Advisory Committee makeup excludes such concerned groups as the apiarists whose bees are slaughtered by insecticides, farm workers who labor in and with the stuff, insectary operators who market natural enemies of pests, consumers whose food and environment are doused with insecticides, and organic farmers who attempt to raise chemical-free crops?

This is how the pesticide mafia has rigged things in its favor in California. What is most disturbing about the California law is that it sets the pattern for the rest of the nation. In other words, there is the disturbing prospect that as examining/licensing legislation is contemplated in other states or at the federal level, eyes will turn to the California prototype and new laws will be formed in its shabby image. Thanks to the politics of pest control, the pest-management advisory profession seems destined to decades of mediocrity, and the environment to a continuing biocidal blight.

Chapter 10

OF APHIDS' KNEES AND BLOODY MARYS

We Americans are fussy people. We shower compulsively, change underwear at the first trickle of sweat, replace our furniture with the seasons, have color-matching telephones and toilets, and insist on consuming bland but beautiful produce. The sad thing about our fastidiousness is that it has been largely foisted on us by Madison Avenue.

Nowhere have we been duped more completely than in our conditioned demand for cosmetically perfect produce. Growers' co-operatives, the food processing industry, and produce retailers have succeeded beyond their wildest dreams in convincing us that we've just got to have impeccable peaches, perfect pears, and spotless spinach.

We are also very fussy about buggy bits that might get into canned, frozen, or bottled produce. And here again our good friends in the food processing and marketing industries have rigged things to make it appear as if there is only the remotest chance of some errant aphid knee or beetle hair finding its way into a tin of succotash or tomato juice, when in fact it is impossible to exclude such tiny specks of protein. Mind you, there are stringent federal and state laws pertaining to produce quality and wholesomeness, which assure the consumer that insofar as insect contamination is concerned, produce is safe and nutritious. But the food processors and some of the growers' co-operatives, in their "concern" over the insectan threat to consumer health and nutrition, insist on outdoing the state and federal bureaucracies. They virtually demand zero bug-knee, bug-bristle, and bug-bite levels, not only in the product as it comes out of the packing house but also in the crop as it is trucked from the fields, and beyond that in the field or orchard itself. In reality they hardly give a damn about consumer health and nutrition; what motivates their drive for zero bug bits is their own competitive game for the prettiest produce in town, a near paranoia over possible lost image in case an insect just might show up in a can of corn, and the associated fear of lawsuit in such an event. The easy escape from these potential horrors is for the processor/retailers to put the onus for bug elimination elsewhere; that is, on the grower; and they have very effective ways of forcing the grower to spray his fields a stipulated number of times each season regardless of insect infestation to "assure" a bug-free crop. A grower's failure to follow the treatment schedule is sufficient basis for contract cancellation. Second, they can and do raise the threat of load dumping at the processing plant in the event the grower has not played the bug-killing game. It is quite easy to find insects or "insect" damage in any load of produce if one really tries. Third, they can simply tell the grower to go seek another contractor if the grower insists on playing the game his way. And even if the poor man finds a new processor, the same rules are invoked.

Games are also played with produce quality depending upon the market economics of a crop. This is routinely done with California navel oranges, which are generally in overproduction so that much of the crop is culled out of the fresh market as "insect-damaged." But on occasion there are brief periods or even entire seasons when navel oranges are in short supply and the price remains high. At these times insect damage (particularly that caused by a thing called the citrus thrips) is largely ignored, as the few available fruits are rushed to market in order to draw a high price. But then, very quickly, when conditions return to normal and there are too many oranges for the fresh market, the price drops and much of the fruit is suddenly found to be "insect-damaged" and diverted to the low-price juice and by-products outlets. Thus, whether the price of his fruit is high or low, the grower is caught in a bind, for there is the constant threat that at any time insect-injury standards will be stringently applied to the oranges. Insect injury is the one fruit-marring factor that he feels he can control, and so whether needed or not he pours on the pesticides in the hope that his oranges will make it to the high-paying fresh-produce market. He is like a dupe playing stud poker: most often his money simply disappears down a pest hole, because, with his crop generally in overproduction, the packing house in one way or another will get much of it into the by-product bin regardless of the amount of spraying.

What emerges from this discussion of cosmetic produce and "zero" insect tolerances is the impression that a tremendous pesticide load goes into the environment to assure that we get our pretty produce or to minimize the off chance of an aphid knee or thrips toe surfacing in somebody's broccoli amandine or bloody mary.[92]

The Environmental Protection Agency, whose charge is environmental quality, became concerned over this situation and decided to investigate it through a contracting agency or individual. I bid for the contract and won it. The contract was most welcome to me, for (i) it permitted the investigation of a matter of considerable concern to consumers, growers, farm workers, public-health officials, pest-control advisers, and environmentalists; (ii) it permitted the establishment of an interdisciplinary study involving representatives from Berkeley's public-health, entomology, resource-economics, and resource-conservation disciplines and the school of law, seemingly a highly desirable program under the mandate of the University's newly established College of Natural Resources; and (iii) it provided support for several graduate students and others associated with the disciplines just cited.

For these reasons I thought that university administration would be delighted with the contract, but I was dead wrong. On the contrary, the contract was an embarrassment to university officialdom, for it generated near hysteria among the food processors and their associates, who in turn brought pressure to bear on the university administration. As a result, I was subjected to continuous harassment from outside and within the University, virtually from the day that the contract was announced.

The trouble started even before I knew that I had been awarded the contract. The first rumble came from a food-processing company executive, who had apparently been tipped off by one of the industry's Washington, D.C., lobbyists. Upon receipt of the word about my contract, the executive phoned me and stated that the food processors intended to complain to the University's vice-president for agricultural sciences about my having received the contract. I asked him what he thought that would accomplish, since I conformed to university rules and regulations in bidding for the contract. He simply ignored this and gruffly asked me how I got the contract and expressed doubts about my competence to conduct the investigation. I responded that EPA apparently considered me to be the best qualified of the bidders. I thought that this took care of the threat to report me to my superiors, only to learn a few days later that a complaint had been delivered to the university vice-president by a food-processing-industry lobbyist, who again asked why I had received the contract and who questioned my competence. To say the least, I was disturbed by this event, not only because industry's representatives had no compunction whatsoever about bringing their bizarre complaint to the attention of university administration but more so because they seemingly anticipated some kind of favorable action on their behalf.[93]

Meanwhile, a member of the vice-president's staff (a former agri-chemical-company executive) took it upon himself to complain to EPA officials about my qualifications to conduct the contract study, and he also denigrated me before certain of my colleagues. I don't really know what he expected to accomplish, because the EPA people ignored him and the only reaction among my colleagues was to call this particular bit of backbiting to my attention.

As I mentioned earlier, the ag colleges include their share of pesticide protagonists, and this apparently explains the actions of the man from the university vice-president's office. It also must have been behind the activities of a colleague from the University's Davis campus, an emeritus top-level administrator who periodically surfaces as a member of those blue-ribbon panels of experts that "review" the pesticide problem. Among other things, he is now a senior scientific adviser to EPA. At any rate, this gentleman tried to blow the whistle on the contract by complaining to EPA's Washington, D.C., headquarters that we had sent out an illegal questionnaire as part of our data-seeking effort. Evidently someone, after meticulously combing the contract and thoroughly analyzing our data-gathering activities for possible irregularities, had passed the information about our "illegal" questionnaire on to the man from Davis, who promptly sounded the alarm. I would not be surprised if the bird dog who did the sleuthing turned out to be a food-processing-industry lawyer who dutifully handed over his pearl to an old friend of agri-business for appropriate action.

Professor Richard Buxbaum, of Berkeley's Earl Warren Law Center, who was a member of the four-man faculty committee advisory to the investigation, conducted an exhaustive search of the statutes to determine the nature of our legal transgression. He did indeed find that there had been a minor legal slip-up but that the onus lay with EPA and not with our investigating team.

The questionnaire we had prepared for submittal to growers concerned their pesticide use patterns under the constraints of cosmetic produce production and "zero" insect tolerances. Grower interviews were vital to our investigation, and since it was very difficult for us Berkeley types (the radical image) to gain entree into grower confidence, we had subcontracted with the Association of Applied Insect Ecologists (AAIE), California's organization of independent pest-control advisers, to conduct the interviews with certain of their client growers, using our questionnaire as a guideline. It was proposed that about forty to fifty growers be interviewed. This is where we got out of line, for somewhere deep in the fine print of the federal statutes there is a clause that states that no U. S. Government agency or contractee thereof may submit a standard set of questions to more than nine members of the public without first obtaining permission from the Office of Management and Budget (OMB).

When this was pointed out to us, we were most contrite in acknowledging our transgression and then asked what could be done to get OMB's permission to carry on with the interviews. We were told (by the EPA officer) that this was a simple matter of filling out a form and submitting it to OMB. He sent us the form, which was in no way simple, and after several days' tedious effort we completed the chore of filling it out and sent it back to Washington. For about ten days, nothing happened. Then I got a phone call from the EPA man, who reported that certain of the questions were not properly answered, and that other of our answers were obscure; would we please clear up these matters. So the form was returned to us and we duly completed the clearing-up process and sent it on back to Washington. Not a peep from EPA or OMB for several weeks. Finally I phoned the EPA man, who said, "Oh, yeah, your petition to OMB; they're going to turn it down." I grunted, "Why?" "Well, they think that AAIE is prejudiced and will run a biased set of interviews because of their pest-control method." I responded, "What method? They use all methods, chemical, biological, cultural, and so forth; they're simply independent pest-control advisers." "Oh! I'll tell OMB that; good-by."

Several weeks went by with not a word from Washington. Meanwhile, our investigation was stalemated while we were waiting for the magic word from OMB to carry on with the interviews. Finally, in desperation I again called my EPA contact. He was very apologetic and told me that he doubted whether OMB would grant us permission to carry on with the interviews. I asked him why, but he simply could not or would not come up with an answer. Finally, a bit angrily, I asked that I be put in contact with someone at OMB to find out just what the hell was going on. He agreed to transmit my request to OMB, and I expected to hear from that agency within a few days. Nothing happened, and so I once again called EPA. When my contact answered the phone, he was obviously very nervous and upset. In fact, he never gave me a chance to ask my questions; he simply said, "Look, Van, please cool it; forget about the interviews; there's a lot of heat coming from OMB. They've sent back word that their man who has over-all control over the EPA budget is in charge of this matter and that he isn't about to deal with anybody; the case is closed." Wow! I was dazzled. Somebody very big had gotten to somebody very big in OMB, and they were putting the screws to EPA. It was obvious that the pesticide mafia and, in particular, its food-processing family didn't want us talking with growers about impeccable peaches and aphid knees. I was a bit flattered, too. Heavy Watergate games were being played to frustrate our little investigation of cosmetic produce. We were obviously on the track of something very rotten. Of course, we knew this from what had already turned up in our investigation, but it was a shock to learn how dirty the game can get when powerful people have something to hide.

Well, I acceded to EPA's wish and cooled it and we never learned OMB's reason for turning us down. What else could I do? Our investigation was stuck with nine grower interviews, and lost some of its punch. But we, nevertheless, completed a rather hard-hitting report, which has the pesticide mafia in a state of hysteria.[94]

In the meantime, while the Washington games were being played, the badgering continued in California. I was called on the carpet by a high university administrator, ostensibly to be reprimanded for using a special-project letterhead on our questionnaire and to be upbraided about the "low" quality of one of our questionnaires. This latter point was raised after a "stonewalling" agricultural commissioner (the kind who plays games with agri-business) complained about the questionnaire to the university vice-president's office.

Next, the college dean got into the act with a letter chiding me about the project co-ordinator, a man he disliked but who had been highly recommended to me by respected colleagues and who performed admirably in his investigative and co-ordinating capacities. The dean also took a swipe at the notorious questionnaire.

So much for the story of harassment. It was not unexpected, and I learned to live with it. But I *am* bitter about one aspect of the cosmetic-produce affair, namely the fact that during the entire ordeal I never received a single spontaneous, heartfelt congratulatory word from anyone in university administration for having sought and received the contract and for setting up the interdisciplinary machinery to effect the investigation. Instead, the administrators scattered from such acknowledgment like a flock of frightened barnyard chickens, pausing only long enough to cluck some words of chastisement.

From this and the hassling I received, it became quite obvious that in the eyes of university administration I had committed a serious transgression in receiving a contract calling for the investigation of a matter of great sensitivity to very powerful people in the agri-business milieu. Indeed, the contract appears to have been an embarrassment to the University. I regret this, not because it brought me into confrontation with university administration but because it provided a very disturbing insight into the way the University is entangled in agri-politics. On the other hand, it is this sort of thing that reinforces my determination to battle against the cancerous encroachment of such politics into my discipline and into the workings of the University of California and similar institutions.

to effect the [illegible]. Indeed the administrators ac-
[illegible] like a flock of frightened
[illegible], [illegible] only some [illegible] to [illegible] some
words of chastisement.

From this and the hearing I received, it became quite obvious that in the eyes of university administration I had
committed a serious transgression in rescinding a contract
calling for the investigation of a matter of great sensitivity
to very powerful people in the agribusiness milieu. Indeed,
the contract appears to have been an embarrassment to the
University. I regret this not because [illegible] me [illegible]
[illegible] with university administration but because it provided a very illuminating insight into the way the University
is entangled in agripolitics. On the other hand, it is this sort
of thing that reinforces my determination to battle against
the cancerous encroachment of such politics into my discipline and into the workings of the University of California
and similar institutions.

Chapter 11

THE RAPE OF EPA*

Several years ago, David Dominick, then the Environmental Protection Agency's assistant administrator for categorical programs invited me to serve as a special consultant to EPA. Dominick told me that the pesticide overload was one of EPA's biggest concerns and said he was convinced that integrated pest management, if widely implemented, would greatly reduce insecticide input into the environment. My job as special consultant would be to explain the integrated-control concept to EPA staffers, so that they, in turn, could better help in its implementation. I unhesitatingly accepted the invitation as a golden opportunity to boost scientific pest control while simultaneously helping to alleviate a serious environmental problem (pesticide pollution).

EPA was full of the juices of youth in those days, and when I got to Washington I found its staffers bursting with a desire to get on with the job of protecting the environment and maintaining its quality. I was caught up in this spirit and poured out my enthusiasm to anyone who would listen. Those were heady days indeed, and it seemed as though nothing but clear sailing lay ahead for EPA and its noble

* This chapter was published largely as it appears here in the January 1976 issue of *Organic Gardening and Farming*.

mission. But then, on one of my Washington visits, I crossed paths with a hardened old bureaucrat who knocked much of the wind out of my sails with the sobering comment: "Forget it, man; I been around here a long time and I'm telling ya, EPA is just like the rest of the agencies set up to protect the public interest: it'll lose its teeth before ya know it. Just watch, in six or seven years it'll be taken over by the people it's supposed to regulate. It always happens that way."

Well, this man was one of the great prophets of my experience. Today the nobly conceived EPA has lost much of its clout and is showing signs of becoming more tabby than tiger. And in no area of responsibility has it suffered greater erosion than in its pesticide registration and regulation function. Indeed, what has transpired can only be described as rape.

The trouble with EPA is that it tried to live up to its mandate. Accordingly, in the pesticide area it took aggressive action and banned such environmentally hazardous insecticides as DDT, aldrin-dieldrin, chlordane, and heptachlor, none of which is critical to the economy or to the public health. But these insecticides *are* of vital interest to the companies that produce and market them. So, too, is the prevailing pest-control system, which is dominated by pesticide marketing imperatives and chemical salesmen. This is where EPA got into trouble, for the American chemical industry wields enormous power in Washington, D.C., and EPA's activities stirred the wrath of this powerful giant. EPA might have escaped heavy punishment if it had stopped its pesticide cancellations with the banning of DDT. But certain of the other hard organochlorines are, if anything, more hazardous, and so the agency quickly banned them, too. This infuriated the chemical giant and turned its thoughts to rape. EPA was raising hell with the pesticide status quo and it had to be stopped and stopped quickly.

Rape is not the usual tactic of the power *mafie* that dominate the contemporary American scene; they much prefer seduction. Seduction is clean, passive, and discreet, and it operates in an aura of respectability. This is what the old bureaucrat of my earlier remarks had in mind when he alluded to the compromising of federal watchdog agencies. But if seduction doesn't work or works too slowly, the violent rip-off will be used without hesitation. This was the case with EPA and its pesticide policies. The virgin watchdog just wouldn't tumble to the seductive overtures of the pesticide mafia, and so it has paid a terrible price.

EPA infuriated the pesticide mafia when it banned DDT. The first sign of rage came in 1974, when EPA Director Russell Train was bullied into permitting the use of DDT against the Douglas-fir tussock moth in the Pacific Northwest (see Chapter 8). But Train is a stubborn or, perhaps, obtuse man, for he ignored or failed to read the real message in the tussock-moth rip-off: "Cool it, Russ! Forget about banning pesticides." Instead, in rapid order he issued decisions banning aldrin-dieldrin, chlordane, and heptachlor. Here he stomped squarely on the toe of the chemical giant, and the giant reacted with uncontained fury, for these insecticides are the big breadwinners of certain of the country's major agri-chemical companies. One of the greatest uses of these materials has been in soil treatment to "control" rootworms in corn. This is an enormous program, in which insecticides are spread over about 50 per cent of the nation's 66 million acres of field corn as an insurance measure against possible damage by rootworms. Insurance treatment for corn rootworm control is an extremely wasteful and environmentally hazardous practice, since in actuality only a small fraction (less than 10 per cent) of each year's crop is economically threatened by rootworms, and can be readily identified.[95] In other words, tens of millions of acres of cornland are annually laced with highly hazardous insecticides to "insure" that a small fraction of the crop will be protected against root-feeding insects. This is an incredibly sloppy way to handle a rather minor insect problem, but it typifies the American way of killing bugs, and the pesticide mafia dearly loves the huge revenues it generates.

Little wonder that Russell Train's cancellation orders provoked the "mafia" into all-out warfare against EPA. Almost immediately following Train's announcements of the aldrin-dieldrin, chlordane, and heptachlor bannings, a fierce barrage of complaints, criticisms, and threats began to pour in on EPA from a multitude of directions. Secretary of Agriculture Earl Butz, the ag mags, the rural media, certain of the urban press, agri-business, grower groups, and corn-belt and corn-pone politicians all rained their grenades on the embattled agency. And Train, again wilting under immense pressure, threw another bone and some more of EPA's teeth to the pesticide mafia; the bone: establishment of the EPA director's Pesticide Policy Advisory Committee. This committee, which can only be described as a tragic joke, has been established to "advise, consult with, and make recommendations to the Administrator of the Environmental Protection Agency on matters of policy relating to his activities and functions under the Federal Insecticide, Fungicide, and Rodenticide Act (FIFRA). Members will be appointed from farm organizations and other pesticide user groups, from the pesticide chemical industry, from private organizations demonstrating an interest in environmental protection, from appropriate state governmental agencies, from among persons known for their expertise in the field of health, and from the general public."

Consisting as it does of members who take sides in the pesticide issue, the Committee is destined, at best, to self-neutralization. This is exactly what the pesticide mafia desired, and in forcing Train to play this card, it took a long stride in getting things back to where it wanted them.

There is potential merit in a pesticide advisory committee, since the biological, ecological, economic, and sociological impacts and implications of pesticides are indeed complex, and such a committee (composed of impartial qualified experts) would be helpful to the decision-making of an overburdened and not necessarily knowledgeable administrator. But the politically inspired nature of the Committee makes it a joke, which from the standpoint of the public interest is a very bad one. Furthermore, Train couldn't have proposed and established a competent, impartial, and effective committee even if he had so desired, because it would not have served the interests of the pesticide mafia and therefore would not have relieved their pressure on EPA.

So today we have the EPA administrator's Pesticide Policy Advisory Committee and, along with it, significant erosion of EPA's pesticide registration and regulation capacities.

But this isn't all there is to the deflowering of EPA. The pesticide mafia is now committed to open rape, and it has tried in the process to break just about every bone in EPA's body. Its main thrust was a bill, HR 8841, amending FIFRA, coauthored by that old friend of the environment Congressman W. R. Poage of Texas (remember him from "The Terrible Tussock Tussle"), which, as passed in a somewhat modified form by the Congress, severely compromises EPA's pesticide-regulating capacity. The major effect of HR 8841 is to give the Secretary of Agriculture veto power over EPA's pesticide regulation and cancellation decisions. In its original version HR 8841 would have given the Secretary outright veto power, but this rip-off was too gross even for the most jaded congressmen, and so a compromise was effected to seemingly soften the U. S. Department of Agriculture's overseeing role. As regards EPA's pesticide regulation proposals, HR 8841 states:

> At least 60 days prior to signing any proposed regulation for publication in the Federal Register, the Administrator shall provide the Secretary of Agriculture

> with a copy of such regulation. If the Secretary comments in writing to the Administrator regarding any such regulation within 30 days after receiving it, the Administrator shall publish in the Federal Register (with the proposed regulation) the comments of the Secretary and the response of the Administrator with regard to the Secretary's comments. If the Secretary does not comment in writing to the Administrator regarding the regulation within 30 days after receiving it, the Administrator may sign such regulation for publication in the Federal Register any time after such 30-day period notwithstanding the foregoing 60-day time requirement.

In Section B of this paragraph on procedure, which concerns "final regulations," the wording is as follows:

> At least 30 days prior to signing any regulation in final form for publication in the Federal Register, the Administrator shall provide the Secretary of Agriculture with a copy of such regulation. If the Secretary comments in writing to the Administrator regarding any such final regulation within 15 days after receiving it, the Administrator shall publish it in the Federal Register (with the final regulation), the comments of the Secretary, if requested by the Secretary, and the response of the Administrator concerning the Secretary's comments. If the Secretary does not comment in writing to the Administrator regarding the regulation within 15 days after receiving it, the Administrator may sign such regulation for publication in the Federal Register at any time after such 15-day period notwithstanding the foregoing 30-day time requirement.

Now, on the surface this stuff seems to be innocent enough, but in actuality, as far as EPA is concerned it is a velvet garrote. There are several reasons for my stating so: In the first place, these provisions will cause delays in any of

EPA's pesticide-regulating actions save those involving grossly apparent and imminent danger to human health (e.g., Kepone®). But, more important, they will almost surely restrict the administrator's actions to those he believes to be overwhelmingly convincing, for the Secretary of Agriculture's counterarguments (especially if he is an agri-chemical-industry champion, as was Earl Butz) will always carry with them the threat of political reprisal if ignored (i.e., there are more agri-oriented congressmen than environmentally concerned ones). And if this seems to be a paranoid observation on my part, I refer to a further provision of HR 8841, which states:

> At such time as the Administrator is required under paragraph (2) of this subsection to provide the Secretary of Agriculture with a copy of proposed regulations and a copy of the final form of regulations, he shall also furnish a copy of such regulations to the Committee on Agriculture of the House of Representatives and the Committee on Agriculture and Forestry of the Senate.

Yes indeed; Mr. Secretary of Agriculture has a garrote around Mr. Administrator's neck, and a whole nest of stilettos backing him up. To be sure, Mr. Administrator is going to be very cautious about promulgating pesticide regulations in that kind of stacked game.

Say a prayer for the birds, bees, and bunnies, and for us folks, too, because the environment is in for a rough time! But this isn't all of the evil contained in HR 8841. The pesticide mafia tried, and in considerable measure succeeded, in seeing to it that there was more than enough toxicant in its EPA mickey finn to knock the agency flat. The kicker in this case was a provision in the bill as it emerged from the House of Representatives, that permitted private applicators (i.e., farmers) to somehow certify themselves as being competent in the use of the more hazardous kinds of pesticides. In the final version of the bill, as modified by the Senate, this "crazy" loophole was slightly plugged by a stipulation that the private applicator must complete a certification form whose adequacy shall be determined by the EPA administrator. But the bill still clearly states that where a private applicator affirms that he has taken a training course, he must not be required to take "any examination to establish competency in the use of the pesticide."

It is difficult to believe that the U. S. House of Representatives could have passed a bill with so dangerous and irresponsible a provision as that which the original version of HR 8841 stipulated for private applicator certification, but in fact it did. The power and influence of the pesticide mafia is indeed frightening!

And even in the final version of HR 8841, the certification standards for private pesticide applicators are disturbingly vague and superficial in light of the fact that the modern synthetic organic pesticides are among the most dangerous and environmentally disruptive chemicals synthesized by man.

Why is the pesticide mafia so anxious to see minimal certification standards for private applicators? The answer is simple: the vast majority of private applicators are farmers who overwhelmingly spray in response to signals received from agri-chemical-company advertisements and pesticide salesmen.[96] If the farmer can be legally sanctioned to apply pesticides without meaningful restriction, then the pesticide industry has an open pipeline through which to pump its toxicants into the environment. HR 8841 goes a long way toward affording the industry this opportunity, and one can rest assured that the pesticide mafia will return again to tinker with the law in an effort to open the chemical floodgates even wider.

As a research entomologist who has pioneered in the development of integrated control, I can only express deep disillusionment, indeed sadness, over what has transpired in the savage rape of EPA. Scientific pest control has suffered a severe setback and may be permanently crippled. But what is most disturbing about this affair is its broader implication of the ugly human condition. We are a vain, rapacious, and foolish species, and as such unable to contain our arrogance and greed in matters bearing on the common social good and environmental integrity. And in this, one has the terrible premonition that as we persist in our foolish ways we will eventually trigger the ultimate tragedy.

Chapter 12

SCIENCE FOR SALE

In 1970 the agri-chemical industry ran up the full hurricane flag as the pesticide tempest gathered force. The national organization apparently decided that the environmentalists were a real threat to the staus quo and that it was time to bring in the heavy artillery and rescue the day. Conferences were held and a battle plan drawn. This plan, which somehow fell into my hands, is too complex and lengthy to detail, but among its facets was a strategy for deep penetration of the scientific societies and the land-grant universities and utilization of those agencies to help tell the "truth" about pesticides.

I am aware of two apparent products of this campaign. One is called CAST, an acronym for Council for Agricultural Science and Technology; the other (now deceased) was called the California Educational Foundation on Agriculture and Food Production (CEFAFP).

The purported goals of each organization seem noble. CAST's purpose is "to increase the effectiveness of agricultural scientists as sources of information for the government and the public on the science and technology of agricultural matters of broad national concern." CEFAFP stated its purpose as "to begin, and continue, a vigorous educational program on the role of chemicals in modern agriculture and on their relationship to the environment and the demands of the public for attractive, safe, and wholesome food."

These are high-sounding objectives, but a peek beneath the scab causes one to wonder just what they mean. Take CAST, for example. I first heard of this group through a letter that its organizers mailed to ag-university administrators in 1972. The letter lamented the lack of input from agricultural interests into the legislative and executive branches of government in matters concerning the impact of agritechnology on the environment. Instead of agriculturalists, the letter complained, consumer groups and persons who do not represent agricultural interests were the principal sources of information on these matters. It further stated that the non-agricultural public, in being concerned about agricultural impact on the environment, received its information all too often from persons with little real understanding.

The letter seemed a reasonable argument for rational inputs by agri-technologists into government and did not arouse my suspicions until, in a late paragraph, there was a suggestion that agricultural scientists take their case to the agri-business industry and solicit financial support.

One wonders whether signals had been sent out that such seed money would be available for the asking. Whatever the case, CAST has had excellent success in getting industry support to help launch its operations. A glance at its list of supporting members reveals such agri-chemical company names as Amchem Products, Inc., American Cyanamid Company, CIBA-GEIGY Corporation, Dow Chemical USA, E. I. du Pont de Nemours & Company, Eli Lilly and Company, Fike Chemicals, Fisons Corp., Montrose Chemical Corp. (famous for its role in the DDT issue), Thompson-Hayward Chemical Company, and Woolfolk Chemical Works, Ltd. Organizations supplying grants in 1974 included Hoffman-La Roche, Inc., Merck & Co., Inc., and Monsanto Company.[97]

In fact, the agri-business firms supply about two thirds of the operating capital that helps CAST inform "the government and the public on the science and technology of agricultural matters," including pesticides such as aldrin-dieldrin, chlordane, heptachlor, and presumably others. For example, during the 1975–76 fiscal year agri-business contributed 64.7 per cent CAST's $116,000 budget.[98]

Above and beyond its strong identification with and financial reliance upon agri-business, what is most disturbing about CAST is the open identification of a number of scientific societies with its operation. In fact, listed on the CAST letterhead, which originates out of the Department of Agronomy at Iowa State University, are the following scientific societies, councils, and associations: American Forage and Grassland Council, American Society for Horticultural Science, American Society of Agronomy, American Society of Animal Science, Association of Official Seed Analysts, Council on Soil Testing and Plant Analysis, Crop Science Society of America, Poultry Science Association, Society of Nematologists, Soil Science Society of America, and Weed Science Society of America. Recently CAST has bagged two additional plums, the influential Entomological Society of America and the Phytopathological Society of America. The hypocrisy of the CAST operation is that it flaunts its "scientific" members on its letterhead but judiciously avoids citing its corporate supporters, who plunk down the bread that makes the thing go. The tragedy of CAST is that it has sucked in thousands of good-guy ag researchers to "represent agricultural interests," while in truth they are primarily serving to enhance corporate greed.

CAST's member "scientific" councils and societies are so genuinely involved with the activities, products, and interests of agri-business, that neither their officers nor the majority of their members appear to discern that they are being used to further corporate interests. I do not mean to imply that these scientists are lacking in intelligence or integrity but, rather, that most, as gentle, narrowly oriented, sincere people, apparently never dream that Machiavellian minds are at work to use them.

As a scientist, I can understand the desire on the part of my peers for the public to know the truth about technical issues, because that is exactly my motivation in speaking out on the pesticide issue. I also know that much distortion has been uttered or published on agri-technical matters. However, the misrepresentations have occurred on both sides of the issue, and it is up to the individual to judge what is right and what is wrong in these cases, and individually to seek a vehicle to express his viewpoint. On the other hand, it is I think completely improper for entire scientific societies to line up in an industry-subsidized club to support pesticides, growth hormones, chemical fertilizers, or what have you and promulgate a pro-agribusiness party line that all is well with agri-chemical practice, while condemning as fools or liars those who dissent. This, in effect, is what the member societies of CAST are doing, and for them to do this is a corruption of the scientific ethic that is both disillusioning and frightening.

In this connection I can relate an interesting anecdote concerning the EPA-supported study of produce standards which I discussed in Chapter 10. When we submitted our draft version of that report to EPA, the Agency, following standard procedure with draft documents, sent it out for comment and criticism to a number of reviewers, including CAST. Typically, these reviews are considered confidential and to be returned by the reviewer to the editor (in this case EPA), who in turn transmits them to the author, who, if he is an astute and experienced scientist, accepts the constructive suggestions and criticisms and improves his manuscript. But in the case of our draft report, CAST simultaneously sent its viciously critical and substantially inaccurate review to the editors of several agri-chemical-industry-supported trade magazines, knowing full well that they would attack the study editorially. And of course they obliged, one labeling the report a "spurious document." In making this move, CAST paid its dues to its corporate keepers, but it also exposed its true nature. I can only hope that from this experience the sincere scientists who contributed to its critique learned a lasting lesson about CAST's "honest," "objective" presentation of agri-technology to government and the public.

The genesis and modus operandi of CEFAFP was as disturbing as that of CAST, and being so close to home, it was a source of deep personal apprehension and revulsion. The prime mover of CEFAFP was the then California Farm Bureau Federation president, Allen Grant, Ronald Reagan's appointee as president of the California Board of Agriculture, ex-officio regent of the University of California, political conservative, farmer-cum-land-developer, and staunch proponent of agri-business and the pesticide status quo. Subsequently, Grant was elected president of the National Farm Bureau Federation, and who knows? if RR had won the presidency, Mr. Grant might have been our Secretary of Agriculture.

One can make his own assumptions as to where Grant got his cue to launch CEFAFP, but it is interesting to note that the foundation got much of its seed money from agri-business (the agri-chemical industry), just as did CAST, and that its originators and/or initial steering committee, in addition to Grant, included such folks as Ivan Smith, lobbyist for the Western Agricultural Chemicals Association; Mel Wierenga, sales executive with Ortho Division, Chevron Chemical Corp.; Robert Woodward, of the Agricultural Chemicals Division, Shell Chemical Co.; Max Sobelman, president of Montrose Chemical Corp., the country's sole DDT manufacturer and perennial bone of contention in the seemingly endless hearings and court cases involving DDT; Dan J. Keating, Stauffer Chemical Company; Dan Niboli, Wilber Ellis Co. (an agri-chemical company); Thomas H. Jukes, a Berkeley medical physicist and as Chevron Chemical Company's Agri-Communicator of the Year[99] one of the nation's most outspoken defenders of the agri-chemical status quo; Hardin B. Jones, a director of Berkeley's Donner Laboratory (physics), a pesticide hardliner; William Hazeltine, mosquito abater and vociferous proponent of DDT; and Jack Pickett, ultraconservative publisher of *California Farmer* and other agri-business-supported journals.

This was some kind of lineup to plan a campaign to "begin and continue a vigorous educational program on the role of chemicals in modern agriculture, on their relationship to the environment and the demand of the public for attractive, safe, and wholesome food." One can hardly doubt that these people had little else in mind than spraying as usual or, better yet, spraying as it was in the good old days.

The formative meetings of CEFAFP were attended by representatives and proponents of the agri-chemical industry and by agriculturists and University of California personnel. I was aware at the time (1970) that an educational foundation on agriculture and food production was in the gestation state and that it would emphasize telling the "truth" about agri-chemicals. I also knew that university personnel were involved, but the names mentioned were not of researchers such as R. F. Smith, C. B. Huffaker, V. M. Stern, H. T. Reynolds, K. S. Hagen, L. A. Falcon, and others deeply concerned with the development of integrated control; instead they were university administrators and extension personnel who had been largely active in rationalizing prevailing pesticide use. The mention of such names as Grant, Jukes, Jones, Hazeltine, Woodward, Wierenga, Ivan Smith, and Jack Pickett in connection with the proposed organization was a further indication that this educational foundation would be dedicated to preserving the status quo.

That there were deep political overtones of a conservative stripe to the group could have been guessed by a perusal of the roster of originators and steering-committee members, who, in addition to Reagan protégé Allen Grant, included Robert Long, a Bank of America vice-president later appointed Under-Secretary of Agriculture by President Nixon, who subsequently served in that capacity in the Ford administration. As mentioned earlier, Long appeared at the 1975 meeting of the Entomological Society of America as a friend of the agri-chemical industry to meat-ax the pesticide-regulating policies of EPA.

I would be wrong to ascribe political motivations to a group, ostensibly concerned with scientific matters, simply because of their political leanings or appointments. But these people have tipped their hand by repeatedly implying that the environmental movement is largely a cover for leftist and radical groups to further their objective of destroying the country's political and economic system.[100]

The following excerpts from the minutes of CEFAFP's formative meeting, held on April 20, 1970, at the California Farm Bureau Federation headquarters, in Berkeley, and attended by agri-chemical industry, agriculture, and University of California representatives, reflect the frightening political overtone of this organization.

> . . . the leftist and radical groups in the U.S. have grasp [sic] the opportunity to use public concern with environmental quality to promote and further their objective of destroying our system of business, industry, and government.

> . . . Professors of liberal and leftist philosophies at universities and colleges across the country seem to be able, without fear of chastisement or loss of promotion, to make irresponsible public statements and claims, while professors having a more conservative philosophy and supported by scientific facts are denied the same "Academic Freedom" by their administrative superiors.

The authors of these incredible remarks then went on to suggest that agriculture, agri-chemical industry, and university (of California) interests should develop an aggressive, positive, factual public-relations and information program regarding agri-chemicals. Evidently they had their own ideas about what constituted facts and how to go about presenting them!

These people dragged the politics of pest control into the gutter on the right-hand side of the street, and in doing so called those who ask questions about the impacts of agri-technology some very dirty names.

CEFAFP never accomplished a thing, and it met a well-deserved end in the autumn of 1974, when it passed the baton to the Council of California Growers, a major agri-business PR, lobbying, and political-pressure group. However, despite its lack of impact and its early demise, CEFAFP still worked a corruptive evil. Most disturbing to me was the success of its instigators in associating this ugly foundling with the University of California. It seems that, like CAST, CEFAFP needed credibility, and what better banner to wave than that of a respected institution such as the University. CAST had a whole string of scientific societies to give it respectability, so CEFAFP apparently set out to get respectability too. The vice-president for agricultural sciences at the University of California accepted membership on the CEFAFP board of directors, as did the prestigious chancellor emeritus of the University's Davis campus.

Furthermore, an old acquaintance of mine, a university agricultural-extension specialist, was among CEFAFP's founding group and was later elected one of its officers. I know this man very well and respect his honesty and personal integrity, although I differ with him on many issues regarding pesticides. I prefer to believe that he was ordered by university brass to partake in the CEFAFP evolution, for I cannot believe that he would willingly run with a pack that considered me and many other persons concerned about pesticide use as being intent upon "destroying our system of business, industry, and government."

As I probed the University's involvement in CEFAFP I became increasingly affected by a feeling of revulsion. At first I had thought that the institution's role was largely symbolic, something forced upon it by the political reality of living with Ronald Reagan and his elitist, pro-establishment credo. But as I studied the documentation, it seemed clear to me that the University was very much a full and willing partner with agriculture and the agri-chemical industry in the evolution of this instrument (CEFAFP) designed to maintain the pesticide status quo and thereby thwart the integrated-control program being developed by many of the University's most dedicated and innovative researchers.

I am probably a hopeless idealist, which is the price I pay for being a scientist. Scientists are molded to seek the truth and tell it. This ethic is the driving force of my life, and I expect it in other scientists. Thus, to me, it is always a shattering emotional experience when I learn of some devious antic by a scientist or a scientific institution. The emotion comes largely as compassion for the errant scientist, who, standing naked and exposed before all his peers, is marked with a brand that survives even beyond the grave: liar, fraud, plagiarist!

My reaction to the role of the University of California in the CEFAFP affair was also emotional, but in this case, involving as it did an institution, it was one of revulsion and sadness. Revulsion because the University played on the side of an organization dedicated to the protection of a vested interest, at the expense of society. In this, the University not only helped cheat society but played a double-dealing game with its own research scientists. There is no place for this sort of thing in a great academic institution.

My sadness came in finally recognizing, after rejecting ample prior hints and warnings, that mother University, whom I have always loved and revered as a virtual saint, had indeed been sleeping around with some rather scruffy dudes.

It is terribly frustrating for someone small and isolated to stand by and witness the corruption of a beloved institution. Mostly, one can only watch helplessly while immensely powerful groups and individuals violate her. The University of California is a great university, which has held up to the forces of corruption reasonably well. Sometimes it bends, but it doesn't break. That I am still around, taking my shots at it, testifies to its resilience. But what bothers me is that sometimes it does bend, and this can only mean that other, less robust institutions scattered over the land do, indeed, cave in. Life must be hell for free-thinking academicians in such violated places.

FRASS

Chapter 13

FREEDOM OF THE PRESS, WELL, SORT OF

Several studies on sources of agricultural pest-control decision-making have revealed that the chemical company fieldman (salesman) and media advertisement collectively dominate grower decision-making.[101] As a major source of revenue, the pesticide industry exercises a strong influence over elements of the media, particularly the trade magazines, the rural press, radio, and television. The grower client is bombarded by this flood of propaganda, which simply overwhelms the technical and popular publications and advice of the U. S. Department of Agriculture and the land-grant universities, and the advice of independent consultants.

A quick perusal of such publications as *Farm Journal, Farm Chemicals, California Farmer, Agrichemical Age,* and *Agri-Fieldman* provides revealing insight into the extent of agri-chemical advertising. For example, the January–February 1976 issue of *Agrichemical Age* had more than half of its forty-seven pages (counting front and back covers) devoted to agri-chemical advertisements. This permits free distribution of the magazine to most of its audience. The following exchange between a delighted ag-university staffer recipient and editor Dick Beeler of *Agrichemical Age*, sums it all up:

> Dear Sirs:
> This is the greatest agricultural publication available. I can't believe it's free.
>
> John J. Reilly
> Assistant Professor
> Blackstone, Virginia

Beeler's response:

> Dear John:
> It's really not free. Our advertisers pay for it and this is a good time to recognize them, not only for that, but for one super, fantastic role they play in serving this nation and its agriculture. How about a nice hand. . . .[102]

Suffice to say, the publishers of the ag mags do not find it expedient to bite the hand that feeds them, and so they present a one-sided story regarding pesticides. In fact, I was once told that chemical-industry representatives informed one editor that the industry would withdraw its advertising if his magazine reported negative aspects of pesticides. Another editor was reportedly warned that advertising would be withdrawn if his magazine published anything favorable about me and my research. Some years ago, a feature writer for California's McClatchy newspaper chain told me that a chemical company withdrew its advertising after a McClatchy newspaper published an article on the adverse effects on wildlife of one of the company's insecticides. These are just incidents of which I am aware. It appears as though the media purveyors of agri-chemical "technology" are just as subject to the coercive whims of their corporate sponsors as are the editors and publishers of magazines and journals that advertise more mundane items such as booze and cigarettes!

And of course, those who suckle from the agri-chemical sow do not hesitate to attack persons who criticize their benefactor. As I mentioned earlier, I have been the target of numerous blasts, being labeled, among other things, a threat to the Republic, a menace to free enterprise, and an incompetent. Rachel Carson was and still is a favorite target of these ag mags, as are William Ruckelshaus (for his DDT decision, the ultimate sin), Russell Train (for his several sins), Charles Wurster (of the Evnironmental Defense Fund), my colleague Ray F. Smith, and others. Recently *Agrichemical Age,* in an editorial entitled "Slandering Agriculture," attacked three University of California economists for simply publishing a carefully researched article that reported that cotton and citrus growers who employ independent pest-control advisers use less pesticide and make more money than do growers who follow conventional control practice.[103] This kind of information is, of course, anathema to the magazine's agri-chemical-industry sponsors. Never mind that it reports on a matter that will benefit the growers, not to mention society and the environment.

In reality these attacks have their positive side: they tell one when he is hitting home. In fact, I have a feeling of accomplishment when I am subjected to ag-mag editorial abuse. Indeed, I have devised an accomplishment-rating system: one paragraph, not so good; two or three paragraphs, something to brag about; a full page—WOW—a barn burner! And if this book ever sees the light of day, I will have to invent a super category, because it will probably evoke enough vilifying editorials to enable me to paper the walls of the family recreation room. What an ego trip that will be!

But while I get my kicks poking fun at the ag mags, they really do disturb me, because of the fierce loyalty that these parasites of agri-business have engendered among growers, agri-researchers, and farm advisers. As mentioned above, a number of these journals are actually throwaways that exist entirely off their advertising revenue and thus preach a pure agri-business party line.

Yet their "public" doesn't seem to realize, or perhaps more accurately, doesn't want to realize that it is being conned (see above exchange between researcher and ag-mag editor). The particular irony here is that the typical grower/reader, who is wasting money on prevailing chemical control practice, probably applauds the ag-mag editorial that labels as slanderous a research report telling him of a more economical way to control pests. At the personal level, on a couple of occasions after having been roasted in one or another of the ag mags for speaking out against the pest-control status quo, I have received letters from seemingly intelligent farm advisers of long acquaintance describing me as a disgrace to the University and a scientific fraud.

The ag-mag editors are, of course, very clever professionals in the game of manipulating reader psychology. They interweave the conservative dogma (i.e., antipathy to bureaucrats, eco-freaks, university radicals) beloved of their clientele with agri-business hucksterism and thereby successfully carve out their livings as scriptive con men. No doubt about it, many an ag-mag editor publisher is the most pernicious kind of parasite, a creature living at the expense of its host and returning nothing of substance.

The pesticide industry owns the ag mags, but its influence doesn't stop there. It can also coerce the giants—such as Time Inc. I know. I witnessed this kind of intimidation firsthand. The tale reminds me of a magazine article I once read about the slaying of a bull moose by wolves. The individual wolf is no match for the moose, and in fact the entire pack can't handle the giant head on. So, as a gang, they harass, hound, run, and rip him to death, especially if at the start he is old, ill, or crippled.

There are wolves in the business world, of course, and they know a cripple when they see one, like that dying giant of the publishing world *Life* magazine. Let me tell you about *Life* and the pesticide industry *lobos*. The episode began with a phone call that I received one day in 1970 from Patricia Hunt, *Life*'s nature editor. With the DDT issue going full blast and the ecology movement in top gear, it occurred to Ms. Hunt that an article on reduced and disciplined pesticide use would be of timely interest. She had heard about our integrated-control studies in California and thought that perhaps a story might develop out of one of our programs. After our discussion she was convinced that an article on integrated control in cotton had merit, and suggested it to the magazine's editorial brass. They agreed. The next move came when John Frook, *Life*'s West Coast editor, came up from Los Angeles to work out the format for the article. Mike Rougier, one of *Life*'s top photographers, was assigned to the project and spent much of the month of August in my laboratory and in the San Joaquin Valley cotton fields doing his thing.

The legwork was essentially finished by the end of August and all the notes and photographs sent to New York. I visited Ms. Hunt in early September, at which time we went over the photographic material and possible captions in what seemed to be the cleaning up of details. Ms. Hunt anticipated that the article would appear within several weeks. But it didn't. In fact, it never did appear. But that's getting ahead of the story.

My reaction to the delay was understandably one of disappointment. My ego wasn't involved, because I was not to be included in the pictorial presentation, nor, for that matter, cited in the accompanying essay. The article was to be on integrated control in cotton, and I looked upon it as a wonderful opportunity to present the case for rational pest control to a wide audience.

Months passed; then, in early 1971, Ms. Hunt phoned me and said that the article was to be published during the spring or summer. A bit later, she sent me a mock-up of the article (which I still have), with its dummy captions. Again nothing happened. And again Ms. Hunt finally contacted me, and for the first time indicated that she felt there was chemical industry pressure to abort the article. I then mentioned that I would be going to Europe in a few weeks and that I could stop off in New York for a day to discuss the article if she believed it would be worthwhile. She thought it was a good idea and said she would talk to her boss (one of *Life*'s senior editors) about it. He agreed that the three of us should get together and thrash things out. So I stopped over in New York and had one of those long, Rockefeller Center business lunches with Ms. Hunt and her boss. Before the first martini arrived, we got down to business. The boss laid it out straight: There *was* pressure from the chemical industry to kill the article. He wanted to hear my story about integrated control firsthand, so that he could reach a final decision as to whether he should proceed with the article or abort it. So, over the course of a couple more martinis, I spilled out the saga of integrated control. When I was finished, he told me that I had convinced him of the validity of the concept and the merits of the cotton article and that he would give it the green light.

As far as I was concerned, that was the end of the story. The article never appeared, and I assume that the agrichemical wolves had their way. Soon thereafter *Life* quietly passed away. I hardly believe that the money *Life* wasted on the aborted cotton article brought it to its knees. It was merely a nip by the wolves. What is certain, though, is that the agri-chemical industry's pressure on the media ranges from the cow-county weeklies to the international opinion giants, which is bad news for public information on rational pest control, and by implication, a lot of other things, too.

Chapter 14

THE SORRIEST LOSER

Some time ago my colleague Louis Falcon was cornered by an employee of a large corporate ranch who related a bitter tale about the loss of his own farm. The man was one of the victims of a bankruptcy wave that struck small farmers in California's San Joaquin Valley during the 1960s. This was a tragic evolution indeed, for some of those losers were dust-bowl refugees of the 1930s (Okies, if you will) who through sheer determination, self-denial, and hard work had re-gained the type of farming enterprise that they had lost thirty years earlier to drought, dust, and depression.

Now economic disaster had visited again, and they were terribly embittered. I don't know whether Lou Falcon's acquaintance was one of those two-time losers, but whatever the case, he was a deeply disillusioned and confused man who blamed his personal disaster on the meddling fools who had denied him DDT with which to combat the cotton bollworm, the pest that did him in. Here he specifically singled me out for criticism because of my stand against DDT in courtroom and legislative hearings. He told Falcon that I shared the blame for his economic disaster, because in helping ban DDT from California I had denied him a chemical tool that was vital to his survival.

He was, of course, dead wrong, for he was in fact a victim of his own ignorance of the ecology of pest control and of the workings of a pest-control advisory system essentially designed to exploit him. In my opinion, there is no one among the victims of the pesticide treadmill more pitiful than the small farmer such as the one just described, who has been nudged down the road to financial ruin by the very chemicals he believes will bring him economic bounty. It is his gullibility and the nature of his victimization that make him so pathetic.

In the treadmill game the victim farmer never has a chance, for things are stacked against him from the very start. In the beginning he receives advice from all quarters that the bugs, weeds, and blights are out to destroy him and that he had better crank up his chemical defenses to protect his livelihood. Roadside billboards, TV and radio commercials, ag-mag advertising, pesticide salesmen, grower neighbors, the feed-store operator, the packing-house fieldman, the county agent, and the official ag-university publications all warn him of the pest peril and exhort him to spray. Pounded as he is by these helter-skelter sources of wisdom, he takes the chemical fix and starts down the road to economic disaster.

The nastiest pesticide treadmill with which I have had personal experience occurred in cotton in the San Joaquin Valley; it was the same one that bankrupted Lou Falcon's farmer friend and many other growers too. The genesis of the problem lay in the indiscriminate spraying of the cotton crop for lygus-bug control, which in turn led to a massive bollworm outbreak. Not only were these infestations devastating, but once the bollworms erupted, very little could be done to contain them, because of their resistance to most of the available poisons and their cryptic habits (they bore into the affected plant parts), which protect them from the insecticides. Nevertheless, the desperate growers sprayed repeatedly in the forlorn hope that they could save their crops. It was a classic pesticide fiasco. Fortunately, an intensive research effort in which Falcon and I and other entomologists participated revealed the cause of the bollworm epidemic and ultimately led to a solution. In a nutshell, the problem derived from totally useless mid-season sprayings for lygus-bug control, which not only entailed a needless expense and pollutive waste of biocides but killed off the natural enemies of the bollworm just at the time when many of the cotton fields were being invaded by hordes of egg-laying moths. With no natural enemies to attack them, the eggs hatched into bollworms, which then munched their way through the cotton, uninhibited by parasites and predators. The result was widespread crop loss.

The bollworm problem was solved by adjusting the lygus-bug sprayings so that they occur only at the time (early in the season) and places (fields with truly threatening infestations) where they are needed. As a result, today the bollworm has virtually vanished as a pest of cotton in the San Joaquin Valley. But before it faded away, the pest victimized numerous growers, and as previously noted, sent some into bankruptcy.

During the course of the epidemic, certain grower casualties came to realize that the basis of their problem was not the bollworm itself but, instead, the insecticides that induced its outbreaks. They decided to take the matter to court to gain restitution, and there again they lost.[104]

I appeared as a witness in two of these cases and observed firsthand the legalistic handling of an ecological rip-off. And here the victims were not birds, bees, or bunnies but good, solid citizens of Middle America who had awakened to the fact that they had been taken to the cleaners by agrichemical companies. But since they based their cases on ecology (i.e., pesticide disruption of the balance of insect populations in crops), they never had a chance. The reason is simple: nothing in the registration or labeling of pesticides requires that their impact on the natural enemies of pests be tested and ultimately noted on the label. In other words, existing law does not recognize the balance of nature, it only directs itself to the pest-killing capacities of the poisons and to their threat to humans, other warm-blooded animals, and certain cherished human possessions. Nothing requires that in the registration and labeling process there be research to determine whether given pesticides can, in fact, aggravate target pest problems and induce destructive secondary pest outbreaks. Nothing on the label warns that the biocide can cause more problems than those that exist. Thus, in using legally registered and labeled pesticides, the grower assumes all the biological and ecological risks, while the producers and sellers of the materials remain totally immune to legal accountability.

The two cases in which I testified involved insecticide-induced bollworm outbreaks in cotton. In the first, the plaintiff, Fabio Banducci, a small Kern County, California, farmer, claimed that an insecticide, Bidrin®, recommended for lygus-bug control by a salesman representing FMC Corporation, induced a bollworm outbreak that severely reduced yield. The key point of contention was that the prescribed insecticide had destroyed the bollworm's natural enemies, thus permitting the pest population to increase explosively. The court entered judgment for the defendant, whose counsel was apparently convincing in his argument that the severe bollworm infestation was simply a natural occurrence and that Fabio's poor farming practices had further contributed to the reduced yield.

The second case involved Hobe Ranches, of Madera County, California, a medium-sized family operation, versus Collier Carbon Co. There the plaintiff alleged that the insecticide Azodrin®, recommended by a salesman to be applied as a single treatment for season-long control of several pests, induced a devastating outbreak of bollworms that severely reduced crop yield. As in the *Banducci* case, the destruction of natural enemes was cited as the key factor contributing to the bollworm outbreak. Judgment was again entered for the defendant, the defense having argued that the bollworm outbreak was a natural event, that some adverse agricultural practice (i.e., poor farming) could have contributed to the reduced yield, and that the responsibility for using Azodrin® fell entirely on the plaintiff.

My role in both cases was to testify on the matter of natural-enemy destruction by Bidrin® and Azodrin®, and on my experience with those materials as bollworm outbreak inducers. My testimony was largely ecological and biological, and as such, it apparently did little or nothing to counter the legalistic presentations of the defense lawyers. Nevertheless, as a biologist-ecologist, knowledgeable of the suspect pesticides and their propensity to cause bollworm outbreaks, I had (and still have) absolutely no doubts about the cause of the Banducci and Hobe Ranches outbreaks.

Fabio Banducci and Hobe Ranches lost a game in which they never had a chance. And what is most ironic about the victimization of these small farmers is that once they realized that they had been snookered, and turned to the courts to plead their case, their agri-chemical industry "friends" didn't hesitate to lash back and "prove" to judge and jury that the plague of worms was a trick of fate combined with lousy farming.

In today's cutthroat agricultural milieu, the small farmer is a vanishing species. (At last reckoning, there were only about sixty-three thousand farms in California, our richest agricultural state.) How tragic it is that as the little guy flounders and sinks into bankruptcy, his own organizations and the grower co-operatives, as well as most of the people who advise him, are in bed with the folks who not only bleed him white but crucify him when he seeks restitution. The small farmer walking to his fate with his eyes wide open is the sorriest loser on the pesticide treadmill!

BOMB DISPOSAL

Chapter 15

INTEGRATED CONTROL—A BETTER WAY TO BATTLE THE BUGS*

The 1975 meeting of the Entomological Society of America was the scene of an interesting comparison between the contrasting insect-control strategies of two of the world's great nations, the People's Republic of China and the United States of America. And from what transpired, it appears as though the Chinese pest-control system has more going for it than does ours. I would like to dwell on this matter a bit, for not only does it cast light on the right and wrong ways to combat insects but also because, if we are willing to read the signals honestly, it gives us considerable insight into what is going wrong with the American way of doing things. There may be something of value in such an exercise.

Insect control in China was described, to an audience of two thousand attending the opening plenary session of the Entomological Society, by a panel of America's leading entomologists who earlier in the year had visited China under the China-U.S. cultural exchange.[105] I know most of the panelists, some intimately, and would characterize them largely as politically moderate Middle Americans. In other words, they had no ax to grind on behalf of China and its Marxist

* This chapter is based in part on an article published in the April 1975 *Organic Gardening and Farming* magazine.

political ideology but reported things as they witnessed and recorded them. From what they had to say, it seems that China's entomologists constantly sift the world's literature and other information sources for relevant techniques, methods, and materials, and integrate them along with their own technical developments into a highly effective national integrated pest-management system. Under this system there is continuous monitoring of pest populations, use of action-precipitating pest-population thresholds (economic thresholds), and the implementation of a variety of tactics, including chemical, cultural, and biological controls, as circumstances dictate.

This program is serving China well. For example: using this pest-control system, China grows 39 per cent of the world's rice, which not only feeds her 900 million citizens but enables her to be a major rice exporter. China also utilizes her pest-management system against disease-transmitting and nuisance insects such as mosquitoes and flies. It is interesting that in mosquito control she employs virtually no DDT, apparently relying instead on reduction of mosquito breeding sources, mosquito exclusion tactics, natural controls, and the judicious use of "safe" insecticides. In this latter connection it is especially noteworthy that China, though producing about one hundred insecticides, relies heavily on seven organophosphates because of their limited hazard to warm-blooded animals. And under her insect-control system, she uses these materials judiciously.

Now let's see how we do things in the U.S.A. Two days after the China report, the Entomological Society heard Assistant Agriculture Secretary Robert Long tell us all about it. On this occasion we were a captive audience, since the convention registration fee included the price of a ticket to the Society's annual awards luncheon, before which industry's spokesman Long performed as "distinguished" guest speaker. In reading the fine print of the meeting program I had earlier discovered that Long's visit to New Orleans was arranged at the behest of the agri-chemical industry. And it didn't take long for him to burst into his expected song as he unleashed a vicious attack on industry's great tormenter, the Environmental Protection Agency. In his speech, Long first chortled over the recently enacted, politically inspired amendments to the Federal Insecticide, Fungicide, and Rodenticide Act (FIFRA), which give USDA considerable veto power over EPA pesticide decisions (see Chapter 11). But then he made it abundantly clear that this was not enough. Despite the FIFRA amendments, Long left little doubt that in his mind EPA still had too much control over the registration and regulation of pesticides, particularly as regards EPA's intentions to seek re-registration of America's fourteen hundred pesticide species and their thirty thousand formulations. Here he ran up the alarm pennant by maintaining that EPA's protocols were so deeply mired in bureaucratic stickum that the agri-chemical industry simply would not make the effort to re-register their materials. In other words, he flatly told us that we were about to lose our thirty thousand pesticides, and he painted a terrifying picture of impending starvation, pestilence, and disease in the wake of this loss.

This rhetoric, as it was intended to do, quite probably frightened the naïve in the crowd while bringing joy to the hearts of Long's chemical-company sponsors. Robert Long, a glib spellbinder, well knew that his prediction of an imminent pesticide wipe-out was complete nonsense. Legal roadblocks and political gamesmanship make this a virtual impossibility. What Long was actually telling us was that the U. S. Department of Agriculture, with powerful political backing, intended to hound EPA into loosening its control over pesticide registration and regulation, to the point where the agri-chemical industry would have things just about as they were in the days before passage of the National Environmental Policy Act. The speech was simply a trial run, with Long using the entomologists to perfect the pitch with which he and other USDA brass planned to bushwhack EPA in forthcoming political jousting.

What he and his sponsors hoped to accomplish, then, was an easing of the way for the American agri-chemical industry to unload its fourteen hundred pesticides in their thirty thousand varieties onto the environment, with USDA bulldozing the path. Fortunately, the 1976 presidential election aborted this plan, which, if it had unfolded, would have permitted the interests of the American chemical industry to transcend environmental quality, public health, and the economic well-being of the farmer and consumer. Madison Avenue would have predominated, while scientific pest control would have remained a fuzzy dream in the minds of a few radical researchers.

But let's return to China. How can she feed, and protect from pestilence, 900 million people, with just a handful of insecticides, while we are led to believe that we must have thousands of poisons or otherwise be overwhelmed by an insect avalanche? Is it that we have a vastly more severe pest problem? I hardly think so. Malaria is nowhere endemic in the United States, but it is in China, as are other horrible, insect-borne diseases. Nor do we have 900 million mouths to feed. What, indeed, has happened is that China has used her intelligence to invoke a national *integrated pest-management strategy,* while our strategy is chemical control dominated by the marketing thrust of the agri-chemical industry. Result: pest-control chaos, and if we care to look about us, we will find that similar chaos characterizes many of the other things that we do.

But it isn't too late to change our ways in pest control or, for that matter, in other aspects of applied technology. As I have mentioned several times, it was a mistake to challenge the insects head on with crude chemical weapons. The bugs are too diverse, adaptable, and prolific to be beaten by such a simple strategy. But we were so dazzled by DDT's great killing efficiency and, perhaps, our cleverness in concocting the stuff, that we ignored the possibility of a bug blacklash and plunged full blast into the chemical "extermination" campaign. And once we had made our move, we were hooked onto an insecticide treadmill just like an addict on junk.

Now, suddenly, in the midst of the nightmare, when our addiction demands heavier doses and more frequent fixes, the chemicals are hard to get and very expensive. Alarmingly, with famine an increasing global concern, many of the chemical eggs in our bug-control basket are no longer effective. The insects, our great rivals for the earthly bounty, are gearing up to march through our gardens, groves, forests, and fields largely immune to our chemical weapons and freed from natural controls. And in the disease area, too, the breakdown is having a disturbing effect, as malaria makes its dreadful resurgence largely because of mosquito resistance to DDT and other insecticides.

The situation would be much more frightening but for a handful of pest-control radicals who never tumbled to the chemical strategy. These are the renegades who quietly worked away on integrated control programs while most in the pest-control arena were on the chemical kick. Though integrated control is still limited in scope, there are enough programs in operation or under development to offer encouragement that there is indeed a better way to battle the bugs.

What Is Integrated Control?

Integrated control is simply rational pest control: the fitting together of information, decision-making criteria, methods, and materials with naturally occurring pest mortality into effective and redeeming pest-management systems.

Under integrated control, natural enemies, cultural practices, resistant crop and livestock varieties, microbial agents, genetic manipulation, messenger chemicals, and yes, even pesticides become mutually augmentative instead of individually operative or even antagonistic, as is often the case under prevailing practice (e.g., insecticides versus natural enemies). An integrated control program entails six basic elements: (1) man, (2) knowledge/information, (3) monitoring, (4) the setting of action levels, (5) methods, and (6) materials.

Man conceives the program and makes it work. *Knowledge* and *information* are used to develop a system and are vital in its day-to-day operation. *Monitoring* is the continuous assessment of the pest-resource system. *Action levels* are the pest densities at which control methods are invoked. *Methods* are the pathways of action taken to manipulate pest populations. *Materials* are the tools of manipulation.

Sounds like what's going on in China, doesn't it!

Integrated control systems are dynamic, involving continuous information gathering and evaluation, which in turn permit flexibility in decision-making, alteration of the pathways of action, and variation in the agents used. It is the pest-control adviser who gives integrated control its dynamism. By constantly "reading" the situation and invoking tactics and materials as conditions dictate, he acts as a surrogate insecticide, "killing" insects with knowledge and information as well as pesticides, pathogens, parasites, and predators. Integrated control's dynamism is a major factor that sets it off from conventional pest control. Thus, though the latter involves some of the same elements, it lacks dynamism in that it is essentially preprogrammed to the prophylactic or therapeutic use of pesticides. In other words, pesticides dominate the system and constitute its rigid backbone.

Where a crop is involved, there is little or no on-going assessment of the crop ecosystem and the dynamic interplay of plant, pests, climate, and natural enemies. This pest-control pattern prevails even in California, our most advanced agro-technology, where over one hundred research entomologists busily at work killing bugs for more than a quarter century have developed fewer than half a dozen valid economic thresholds for the hundreds of pest species. A perusal of the stack of official University of California pest-control recommendations reveals the following kinds of pest-control action criteria:

when damaging plants
when present
when damage occurs
when they first appear
when colonies easily found
when abundant
when needed
early season
when present in large numbers before damage occurs
anytime when present
early, mid, and late season
on small plants as needed
when present and injuring the plants
when feeding on the pods
throughout the season
when infestation spotty
when plants are three feet tall.

What this long menu of senseless gobbledygook implies is that in California the insecticide folks have a wide-open field in which to hustle their chemicals, and this they do with greater success than anywhere else in the world.

Under the prevailing chemical control strategy, there is virtually no flexibility in decision-making, particularly as regards alternative pathways of action. The game plan is set at the start and it is stubbornly followed. Result, the familiar case of the fruit grower who year after year automatically sprays his orchard a dozen times or more with the calendar as his main decision-making guide. Or the cotton grower who typically sprays when a chemical-company fieldman drops around and tells him that a few stinkbugs, bollworms, or army worms are showing up in the south forty.

In conventional pest control, one turns on the chemical switch, sits back, and lets the insecticides do the job. It is the lazy man's approach, which characterizes so many aspects of modern life and for which society and the environment pay dearly. A measure of this cost can be gained from a brief analysis of pest control in California.

California's pest control is locked to chemical pesticides. The state is the country's greatest user of these materials, and as stated earlier, receives about 5 per cent of the world's pesticide load. It appears that along with its primacy in smog and earthquakes, California has another distinction: leadership in pesticide pollution. Little wonder! More than fourteen hundred chemical-company fieldmen (salesmen) prowl the state, servicing the prevailing pest-control system. They assure a sustained chemical blizzard as well as a fat market for the agri-chemical industry. And at what a cost! Not only does this horde of hustling polluters dump hundreds of tons of unneeded pesticides into the environment, but in the bargain they annually cost California's economy about $50 million to support their huckstering. The chemical companies and many of the major pesticide users (growers, mosquito abaters, forest pest controllers, and pest-control operators) don't pay the bill, they simply pass it on to the consumer, who doubles as taxpayer. But the story doesn't end with money needlessly spent; there are also ecological and social impacts (see Chapter 3), which add immensely to the cost of the prevailing chemical control strategy.

What I have just described for California pretty much

characterizes pest control for the United States in general, and for that matter, other of the world's modern agritechnologies. Chemical pest control, like so many of our modern practices, is a technology gone wild under the merchandising imperative. And as with our other excesses, this rampant technology must be brought under rein if irreparable damage is to be avoided. I am convinced that we pest-control researchers (particularly entomologists) have the capacity to turn things around through integrated control, and perhaps coincidentally establish a model of technological responsibility for other disciplines. But first it is perhaps best to summarize several integrated-control programs so as to provide insight into the operational mechanics of the strategy and into the benefits it brings.

Integrated Control of Mosquitoes in Marin County, California

Marin County is basically a posh bedroom and weekend retreat for people who do their business in San Francisco. As such, it is populated by a mixed bag of intellectuals, free spirits, artists, poets, filthy rich, potheads, nature worshipers, drifters, and a few just plain folks. In other words, it is a very sophisticated place. Quite appropriately, then, Dr. Allen D. Telford and his colleagues in the Marin County Mosquito Abatement District have developed a mosquito integrated-control program that ranks with the most sophisticated in the country. The program, which involves population monitoring, reduced pesticide use (why spray mosquitoes that don't bite anyone?), and breeding-place management, has resulted in a more than 90 per cent reduction in spraying while effecting an over-all reduction in the mosquito problem.[106]

The most striking element of this program has been the management of mosquitoes in the two-thousand-acre Petaluma Marsh. At one time, this wetland was a major mosquito producer, contributing to both urban and livestock problems. As a consequence, most of the marsh was sprayed with deadly parathion by aircraft five times a year. Dr. Telford and his colleagues shrewdly deduced that the mosquito source was not the marsh's maze of sloughs and channels, which are subject to tidal flushing, but, instead, "potholes" mostly created by dummy bombs dropped during World War II when the area was a practice bombing range. Other human activities created the remaining mosquito-breeding sites. The potholes were not subject to tidal flushing; thus, after flood tides many retained water, which stagnated and became ideal mosquito-breeding habitat. So the Marin County entomologists acquired a ditching machine and developed a pothole drainage system that permits tidal flushing. The program has been so successful that there are only a few, as yet undrained, holes that still require hand spraying. Today no aircraft drone over Petaluma Marsh excreting their lethal organophosphate insecticide onto the teeming life system. Yet the mosquito problem has disappeared from nearby communities, and dairymen operating adjacent to the marsh have told Telford that their herds are free of tormenting mosquito swarms for the first time in memory.

Thus, through an imaginative, integrated-control effort, California's Marin County and neighboring Sonoma County have realized substantial ecological, economic, and social benefit.

Integrated Control of Street-tree Pests in Berkeley, California

Several years ago, complaints by Berkeley citizens concerning the city's tree-spraying program brought Park and Recreation Department officials together with University of California entomologists to plan an integrated control program to minimize insecticide use.[107] The program, largely developed by William and Helga Olkowski, has been a milestone in urban pest management. It's about time, too, since in terms of volume used urban pesticide use essentially equals that in agriculture, and with millions of people in close contact with the materials there is probably more of a human health hazard in the cities than down on the farm.

What the Olkowskis found when they first probed the Berkeley pest-control system was utter chaos. The city's pest-management people were well informed about tree and plant identities but they knew virtually nothing about pests: their identities, activities, and interactions with natural enemies. Consequently it never entered the city workers' minds to seek long-term biological, mechanical, or cultural solutions to pest problems. Instead, typically when a citizen complaint came in concerning a sick or bugged tree, a city crew roared out with its old reliable spray rig and doused the "suffering" tree and more often than not, just for good measure, all the rest of the trees along the block too. An awful lot of unnecessary pesticide spraying!

So the first thing the Olkowskis did was to persuade the city people to change their action pattern from one of automatic spraying to one of first inspecting the tree or trees to determine whether there was even a pest problem at all. Frequently the trees were just old and "tired," or suffering from poor moisture conditions, soil compaction, or malnutrition. And even where insects were found, their damage was often inconsequential or at most secondary to other misfortunes the trees were suffering.

Things are different in the city insofar as insect injury to trees is concerned. The insects rarely do permanent damage, and so, unlike their country cousins, the city bugs are more an aesthetic problem than an economic one. People just don't like the sight of them or they don't like them riddling the leaves on the trees out in front of the house. So they call up the city Park and Recreation folks, and out comes the spray rig.

The Olkowskis met this problem by inventing the aesthetic-injury level (the point at which a citizen can no longer stand the sight or evidence of insects), and then educated the Berkeley pest-management people, and through them the citizenry, to raise their aesthetic boiling point; that is, to tolerate quite a few more insects than before. This took care of much of the unnecessary spraying.

A major source of citizen complaints in Berkeley is the mess created by the honeydew excreted by aphids, particularly species feeding on linden, elm, and oak trees. The linden and elm aphid problems were largely solved by biological control effected by parasitic wasps imported from Europe. Imported parasites also helped with the oak aphids, and when and where they didn't do the job, plain-water and water-and-soap-solution sprays were substituted for the organophosphate insecticide previously used. Ant control with sticky bands around the tree trunks also helped reduce the aphid problem. Aphids are ant cows, providing their keepers with honeydew. If the ants can't get to the aphids to tend them and harvest their "milk," the aphid colonies suffer predation and parasitization and decline in vigor. The sticky bands kept the keepers from the cows, and the aphid problem declined.

In the case of the bothersome California oak moth, the selective microbial insecticide *Bacillus thuringiensis* proved to be a more than adequate substitute for the broadly toxic chemical insecticides in previous use.

The Berkeley integrated-control program, which involves about thirty thousand trees on one hundred twenty acres, has been an outstanding success and a model of its kind. It has had spectacular effects, including the virtual elimination of synthetic organic insecticide use and a savings of about twenty two thousand a year to Berkeley's Park and Recreation Department (see Table 1). In a recent public statement, Mr. Grayson Mosher, retired Berkeley city parks supervisor, remarked that his association with the integrated-control program was the most rewarding experience of his entire professional career. Currently, similar pro-

Table 1. Insecticides used during period 1969–75 on shade trees by the Department of Recreation and Parks, Berkeley, California. Adapted from Olkowski et al., 1976 (Notes and References Item 107).

Insecticides	Amounts used*						
	1969	1970	1971	1972	1973	1974	1975
DDT	65	0	0	0	0	0	0
diazinon	16	6	2.5	1	0.375	0	0
dimethoate	9	7	3.7	2	0	0	0
malathion	12	7	1.9	0	0	0	0
meta-demeton	2	0	0	0	0	0	0
dicofol	2	0	0	0	0	0	0
alkyl aryl sulfite	0	3	0	0	0	0	0
*Bacillus thuringiensis***	0	30	3.0	21	2	0.5	7.5
chlordane	0.5	0.5	0.5	0	0	0	0
lead arsenate	200 lbs	0	0	0	0	0	0
carbaryl	60 lbs	20 lbs	0	0	0	0	0
lindane	0	1	0	0	0	0	0

* All amounts are in gallons except as noted.
** A biological insecticide.

grams are under development in the cities of San Jose, Palo Alto, Modesto, and Davis.

Integrated Control of Spider Mites in Washington State Apple Orchards

One of the most spectacular temporary "victims" of DDT was the codling moth, the legendary worm in the apple. However, as time passed, the worm was able to adjust to DDT, forcing the use of a succession of materials to combat it. But the worm is not the only apple pest. Unfortunately the crop is plagued by a variety of insect and insect-like species as well as several severe diseases. In fact, apple is one of our most pest-plagued and hence heavily sprayed crops.

Apple growers managed to stay ahead of the codling moth and other bugs and diseases, with multiple sprays, until one pest group, the spider mites, began to "handle" the available pesticides. The aggravated spider-mite problem was a classic example of an induced secondary-outbreak pest that became resistant to virtually all available poisons. In many areas the problem verged on intractability and demanded a solution if economic disaster was to be avoided. In Washington's Yakima, Wenatchee, and Okanogen valleys integrated control was the chosen path to salvation.

Development of the Washington program has been largely the result of the "quiet genius" of Dr. Stanley Hoyt, of Washington State University.[108] Hoyt's program, which is one of the world's classics in rational pest control, has served as a prototype for similar programs in apple-growing areas of the Middle West and the Northeast.

The Washington program is basically oriented to the protection of a predatory mite, *Metaseiulus occidentalis*, which is the key natural enemy of the pest spider mites. In studying the predator, Dr. Hoyt found that among other things, it was resistant to a variety of pesticides. This important finding played a key role in the integrated control program, since it permitted the use of materials and dosages that are effective against the target pests but do not interfere with the predator. The program, as is typically the case with integrated control systems, employs continuous pest and natural-enemy monitoring so that population trends and pest/predator ratios are always known. This information is indispensable to control decision-making.

There is a further interesting wrinkle to the program in that it utilizes a plant-feeding spider-mite species, the apple rust mite, formerly sprayed as a pest, to sustain the predatory mites during periods of pest-species scarcity. Hoyt and his coresearchers found that the apple rust mite, though often abundant, rarely causes sufficient damage to merit control measures. They also found that it is an important food source for the predator *Metaseiulus*. With this knowledge they stopped treatments for the rust mite, thereby directly reducing control costs and enhancing populations of the predator to the extent that it provided highly effective control of the two pest species, the McDaniel mite and the European red mite.

The Washington spider mite integrated-control program is employed on over forty thousand acres and has been a

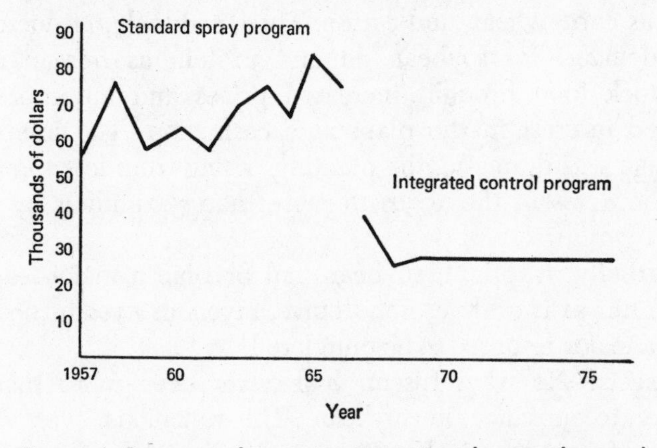

Figure 4. Pest-control costs in approximately one thousand acres of apple in Washington prior to and following adoption of an integrated control program for spider mites. Data from S. C. Hoyt, and L. E. Caltagirone, Chapter 18, page 410, of *Biological Control*, C. B. Huffaker (ed.), Plenum Press, 1971, and from personal communication with S. C. Hoyt. The depicted costs for the years 1970 to 1975 are based on Dr. Hoyt's statement that they varied from about twenty-thousand dollars to twenty-five thousand dollars per year during that period. Hoyt further stated that the cost reduction for spray materials in the integrated control program has averaged about 70 per cent.

substantial success. Not only have control costs been lowered, but in some orchards spider-mite sprayings have been entirely eliminated. What's more, the selective and diminished use of pesticides has permitted the build-up of natural enemies of certain of the other apple pests, reducing the need for chemical control of those species. Where the program has been faithfully employed, pest-control costs have been cut by more than 50 per cent (see Figure 4).

Integrated Control of Soybean Pests

For many years soybean languished as one of our moderately important crops, being far overshadowed by such giants as corn, wheat, and cotton. Then, suddenly, an increasing demand for soybean oil and protein as human and livestock food brought increased prices and an unprecedented interest in the plant as a cash crop. The jump in acreage was dramatic, the plantings rising from less than 20 million acres in the 1950s to more than 50 million by the mid-1970s.

Virtually overnight, soybean had become a major-league crop. But as it grew in importance, ravenous eyes turned to it as a major resource to be plundered.

Insects? No way! Insects had never been more than a moderate nuisance in soybean. The rapacious eyes were those of agri-chemical-company executives who saw a 50-million-acre market ready to be milked. But before they could start a serious pesticide-dumping exercise, other men had made a countermove. Those were the integrated-control-oriented entomologists in the soybean belt, such men as Dale Newsom and Walter Rudd in Louisiana, Sam Turnipseed in South Carolina, Marcus Kogan in Illinois, and Will Whitcomb in Florida. These people perceived what was in the minds of the agri-chemical raptors and shut them off at the spray nozzle with an integrated control program that is one of the finest in existence.

The soybean program is unique, because it was developed in a crop that was free of serious insect problems and was devised to keep things that way, by preventing even the beginnings of a pesticide treadmill. By contrast, virtually all other integrated control programs now in effect were the products of crisis.

Again, in the soybean system as in the others described here, all the components of the classic integrated control program have been brought to bear.

Men were involved in devising the program, and men make it work. In practice there is regular scouting (monitoring of the fields) to assess crop development, pest-population levels, pest-caused damage, and the levels and performance of natural enemies. Economic damage thresholds involving both defoliators and pod-feeding insects are employed as the basis for insecticide use decisions. Where chemical control is required, selective insecticides are used at minimum rates. Methods have been developed to enhance natural enemies not only by use of selective insecticides but through cultural practices and the release of insectary-produced parasites of the Mexican bean beetle.

A number of agronomic and cultural practices have also been invoked to limit pest-insect populations and injury. For example, early-maturing soybean varieties are used in some areas to reduce populations of caterpillar pests. Another practice is to select planting dates so that the crop is unattractive to such pests as the corn earworm (yes, it attacks soybean, too!) at the time of the pest's peak flight. Where soybeans are late-planted (after wheat), the crop is seeded in narrow rows so that the plants quickly form a canopy which discourages corn-earworm attack. In another crop-management wrinkle, rows of snap beans are planted as a trap crop for the Mexican bean beetle, which is then either sprayed on this limited vegetation or assaulted by mass-released insectary-produced parasites.

The soybean integrated control program is imaginative

and effective and is in widespread practice in the Midwest and the South. It is a quiet triumph because it preserved a condition of nearly perfect natural balance. It is difficult to measure the benefit of this kind of program, because, like preventive medicine, it has forestalled an epidemic—in this case, a massive pesticide treadmill in a giant crop. But if we look at cotton and reflect upon what has happened to it globally, we can perhaps get some idea of what might have been in store for soybean had it not been for the foresight, energy, and imagination of a handful of researchers.

Integrated Control of Citrus Pests in Tulare County, California

Citrus is an enormous industry in California, and Tulare County is the state's greatest producer, with approximately fifty-five thousand acres under cultivation. The crop is attacked by a half dozen major pest insects and several minor ones, whose chemical control adds an enormous burden to the grower's production costs. Fortunately, a team of researchers that includes my Berkeley colleague Charles E. Kennett, Dr. Daniel Moreno of the USDA Agricultural Research Service, Tulare County Farm Adviser Donald Flaherty, and private pest-control consultants James Stewart and James Gorden, have developed an elegant integrated control program that is now bringing efficient pest management at lowered cost to many growers.

The program that this research team has developed involves all the components of a classic integrated control system: (1) man, (2) knowledge/information, (3) monitoring, (4) decision-making criteria (economic thresholds), (5) methods, and (6) materials and agents.

Under this program the citrus groves are intensively monitored the year round by the pest-control consultants. In given seasons certain pests are monitored more intensely than others, but no month goes by in which the pulse of the crop is not being taken. In this monitoring process several techniques are used: pheromone trapping for male California red scale, fruit inspection for citrus thrips, leaf examination for citricola scale, shaking of branches into collecting nets for cutworm population assessment, visual counts for leaf rollers and katydids. The counts gathered by these methods are then compared to specific economic thresholds that have been developed for each of the major pest species. Where the data reveal damaging or threatening infestations, control measures are indicated. Here again the researchers have shown foresight, for they have developed selective chemical controls for virtually all the pests. Thus, when a spray is called for, they can plug in a material that effectively knocks down the given pest without unduly disrupting the natural enemies of the other noxious species. For example, with the California red scale, use of the selective carbaryl-oil mixture in conjunction with population-growth assessment by pheromone trapping (the traps are baited with live females, which draw in males and thereby give an index of over-all population density and growth) results in control that lasts for from two to three years, while in the programmed prophylactic spraying used by many growers in the area, insecticides must be applied once a year.

Under the integrated control program, natural enemies of both major and minor pests are substantially protected. This permits them to take maximum toll of their prey and helps lengthen the gap between sprayings. If it were not for the cosmetic treatments required by the packing houses for citrus thrips control, insecticide spraying for citrus pests in Tulare County would be at an even lower level of intensity under the integrated control program. But despite this forced spraying for a cosmetic pest, the integrated control program keeps the pesticide load at a minimum by utilizing selective insecticides only when thrips population counts indicate the need for spraying.

Consultants Stewart and Gorden employ the integrated

control program on about 10 per cent of Tulare County's citrus acreage. Yet, despite paying a consulting fee, the per acre pest-control costs of the Stewart and Gorden clients are less than those of the average Tulare County citrus grower and their produce stands up with the best at the packing house. A recent study by University of California agricultural economists Darwin Hall, Richard Norgaard, and Pamela True indicates that not only do the citrus orchardists using the integrated control system apply less pesticide and thus save money but they also appear to produce a better crop than do their heavily spraying neighbors[109] (see Table 2).

Table 2. Average insecticide costs and dollar yields per acre for groups of cotton and citrus growers utilizing conventional chemical control versus those for growers utilizing integrated control (independent pest-management advisers). The San Joaquin Valley, California, 1970 and 1971.
From Hall et al. (Notes and References Item 109).

Cost + Yields	Two-year Average Insecticide Costs and Dollar Yields/Acre			
	Cotton		Citrus	
	Conventional Control	Integrated Control	Conventional Control	Integrated Control
Insecticide Costs	$ 11.97	$ 4.94	$ 42.35	$ 20.53
Dollar Yields	247.00	270.20	502.85	515.80

Integrated Control of Cotton Insects in the San Joaquin Valley, California

The cotton growers of the San Joaquin Valley have traditionally spent about $25 million a year to protect their $500-million-plus crop from "ravaging" insect pests. But as is so often the case, much of this expense is unnecessary. My own studies, conducted jointly with Dr. Louis Falcon, as well as the research of other colleagues, indicate that there is a striking excess use of insecticides in San Joaquin Valley cotton, amounting to perhaps as much as 50 per cent over what is needed. This adds up to a staggering waste of money and materials and shameful environmental pollution.

Fortunately, we have developed an integrated control program which, when widely adopted, will reduce pesticide use in cotton to the effective minimum while bringing economic, ecological, and social benefit to the surrounding area. The program, one of the most sophisticated of those discussed here, has been implemented in a substantial percentage of the San Joaquin Valley's 1 million acres of cotton.

Basically, it entails continuous assessment of cotton plant growth and fruiting performance in relation to climate, insect populations, irrigation, fertilization, cultivation, and to some extent, disease. The integrated-pest-management specialists visit the fields at frequent intervals from mid-May until the end of September. During these visits they assess plant growth, note fruiting performance, measure insect populations (both noxious and beneficial), and record insect injury. Meanwhile, the cotton performance in individual fields is plotted against an optimum performance chart for the variety under cultivation (Acala). Deviations from optimum performance call for an assessment of the spectrum of factors (e.g., irrigation, climate, insects, fertilizer), that could possibly be causing poor performance. This takes much of the guesswork out of decision-making, particularly as regards the role of insects and insect control. Too often insects are blamed for crop loss when, in fact, bad agricultural practice or adverse climate is the cause. For example, tons of insecticides are poured onto San Joaquin Valley cotton fields for "bug damage" that is, in fact, injury resulting from poor irrigation, poor cultivation practices, or temperature extremes. Under the cotton integrated control program the guesswork over insect injury is even further reduced by the employment of rather precise economic thresholds (the point at which insect damage is sufficient to justify artificial control measures) and an understanding of the high-hazard period of the single potentially serious major pest insect, the lygus bug. Where the cotton integrated control program has been put into practice, insecticide use has been held to a minimum and crop yields and quality have been maintained at optimum levels. For example, one of the largest corporate ranches in the San Joaquin Valley, which essentially follows the program just described, has reduced its insecticide use and per acre control costs by more than 50 per cent, while maintaining high yields. Small growers, too, have benefited in a similar way. Over all, one of the greatest direct benefits of the program has been the virtual disappearance of the bollworm and other caterpillar species, which historically had occurred as secondary-outbreak pests in the wake of excessive and poorly timed insecticide spraying for lygus-bug control. In fact, the bollworm has become so scarce that Dr. Andrew Gutierrez was forced to send one of his Ph.D. students to Mexico to find sufficient high populations upon which to conduct the final experiments of his desertation study.

An even greater sophistication seems possible for integrated control, over what has already been attained. I have in mind the elegant programs that will evolve out of mathematical models being devised for the growth and development of various crops and other resource-providing systems. A model is simply a mathematical description of how the components of an ecosystem fit together. In an agro-ecosystem, a model enables researchers to examine such diverse factors as fruit growth, moisture balance, nitrogen use, etc., as influenced by weather. Basically, what is being accomplished in the development of these models is a very thorough understanding of the crop plant or other resource species, and those factors that affect its performance. With agricultural crops, the first step is to model the plant itself. Current programs simulate the growth of the crop as influenced by weather, various cultural practices, and pests. For example, the University of California's Dr. Andrew Gutierrez and his associates, in developing the simulation for cotton, first converted an available single-plant model to a cotton-population model (i.e., a crop model) and have since methodically added subroutines for such things as temperature, sunlight, moisture, fertilizer, insects, disease, etc. The simulation as it evolves is repeatedly tested against actual crop performance in the field.

The models are only as good as the science that goes into them. Thus, in order to have credibility they force us to do very good biology both as regards the crop plant itself and the organisms that affect it, such as insects, diseases, weeds, etc. Furthermore, additional good biology is required to gain an understanding of the roles of other biotic agents in the system, such as the predators, parasites, and diseases of the plant-feeding forms. Gutierrez and his group working in cotton have already accumulated the best store of biological information on such pests as the lygus bug, bollworm, beet army worm, and pink bollworm of which I am aware.

It is to be emphasized that a model does not control pests but, instead, serves to give us a very clear understanding of a crop production system and the roles of various climatic, biological, and agricultural parameters in crop performance. The model's obvious benefits are

(1) The pinpointing of real problem areas, which takes most of the guesswork out of assessing the roles of the various production-influencing factors in the system. For exam-

ple, the cotton model shows that insects have little effect on San Joaquin Valley cotton production, whereas weather can be enormously important. Interestingly, Dr. Gutierrez has told me that in cotton the dreaded lygus bug is in fact most often a beneficial insect, because in pruning off excess fruit it permits the plant to fatten up those bolls it can support and thereby increases yield.

(2) By identifying problem areas, the model permits the establishment of meaningful research priorities and therefore the efficient allocation of manpower and funds. Thus, the developing alfalfa model and its economic submodel for weevils show that long-term dependence on chemical control of the weevil will be uneconomical. This tells us that research emphasis should be placed on other pest-management tactics, such as biological control, development of weevil-resistant alfalfas, and cultural manipulations.

(3) The models have some short-term predictive capacity relating to both pest population trends and crop yield or quality. Long-term predictive capacity is not envisioned, because the models are temperature-driven, and we simply cannot predict weather with precision. However, with pests such as the alfalfa weevil, a population-prediction capability of several weeks is possible and could be an extremely useful factor in control decision-making for such a single-generation pest. Mini-models of codling moth in apple and pear have already enabled orchardists to more precisely time their sprays. This not only results in greater control efficiency by permits use of lower doses of insecticide.

(4) The models can provide extremely valuable insight into the injury potential of such species as the greatly feared pink bollworm of cotton, which in California is a constant threat to invade the San Joaquin Valley. Dr. Gutierrez informs me that the data being developed for the pink bollworm strongly indicate that the pest does not have the capacity to develop to significantly injurious status in the San

Joaquin Valley. This has important immediate economic implications, since the San Joaquin Valley cotton industry is now expending about $1.25 million per year on a pink-bollworm eradication program.

Well, what does all this modeling business mean as regards the future of pest-control advisement? I think that it will have a profound effect, for it will help to turn pest-control advisement into a respectable technology comparable in most ways to other high-grade technologies. As this evolution occurs, much of the money that society now spends on pesticide-company salesmen-fieldmen will be redirected to support a cadre of agro-technologists (many in private practice) who will be the practitioners of integrated pest management. These practitioners will not simply be bug killers but, instead, crop-production specialists who will have at their command the background knowledge, informational inputs, and decision-making criteria that will enable them to manage pests economically, efficiently, and safely while orchestrating the other components of the crop production system.

As limited or as speculative as they may be, these models all entail or envisage the utilization of data-gathering, interpreting, and decision-making personnel. These are, or will be, technically trained people who will (it is to be hoped) be the pioneers of the integrated pest-management cadre.

Thoughts About the Future

I have discussed several highly effective operational integrated control programs. The emphasis has been on programs in California, but I should note that there are similarly effective programs elsewhere in a variety of crops including cotton, apple, alfalfa, soybean, and tobacco in such other states as Arizona, Arkansas, Florida, Illinois, Indiana, Louisiana, Michigan, Mississippi, New York, North Carolina, Pennsylvania, Texas, and Washington. Furthermore, there are programs in a variety of crops in foreign countries, too; for example, Australia, Israel, India, Sri Lanka, Malaysia, Peru, England, the Netherlands, Switzerland, and the Soviet Union, as well as China. These programs share several things in common. *All* involve continuous ecosystem assessment. *All* have produced efficient pest control. *All* have resulted in substantial reductions in pesticide use. Finally, most have effected monetary savings. And while doing these things, *all* have maintained or even increased crop yields and quality as well as public health standards.

The striking reduction in pesticide use or pest-control cost under integrated control is illustrated by the following programs, all of which have, at equal or lower cost, provided equivalent or even better control than the programs they replaced.[110]

Crop/Resource	*Locality*	*Reduction*
Apple	Washington State	a greater than 50 per cent cost reduction
Pear	Sacramento Cty., Calif. (1976)	about an 85 per cent reduction in worm-control costs
	Lake Cty., Calif. (1976)	about a 50 per cent cost reduction in worm control
Cotton	California (San Joaquin Valley)	about a 50 per cent reduction in insecticide use
Cotton	Arkansas (100-square-mile area)	an 80–90 per cent reduction in insecticide use
Cotton	Texas (trans Pecos area)	insecticides virtually eliminated

Citrus	California	about a 50 per cent reduction in insecticide use
Mosquitoes	Marin Cty., Calif.	nearly an 80 per cent reduction in insecticide use
Highway vegetation	California (1969–71)	nearly an 80 per cent reduction in insecticide use
Street trees	Berkeley, Calif.	a greater than 90 per cent reduction in insecticide use
Tomato	California	about a 50 per cent reduction in insecticide use

It would thus seem that integrated control is indeed a better way to battle the bugs. If this is true, then why hasn't there been a swift and sweeping shift to this strategy? Why is it that in California, for example, integrated control is practiced in only 10 per cent of the cotton fields, in an even smaller fraction of the deciduous fruit acreage, and in but a handful of communities and mosquito-abatement districts? Why hasn't it spread rapidly to other areas?

The answer is complex and touches on several aspects of human nature, including the familiar arrogance, foolishness, and greed. Among these, greed is paramount. Powerful forces have a vested interest in the pest-control status quo. The agri-chemical industry, in particular, is not about to stand aside while the pest-control baton is ripped from its hand and, along with it, profitable license to overload the environment with insecticides. To it, integrated control is anathema, and it is fighting the curse with all of the political and media muscle it can muster. Recently in an unprecedented power play, the chemical industry rallied a hooting, stomping mob of sixteen hundred agri-business supporters to

Sacramento, where, in a lynch-mob atmosphere, it bullied a California Department of Food and Agriculture (CDF&A) panel hearing an integrated-pest-management petition submitted by the Environmental Defense Fund. The industry was aroused to near hysteria by EDF's proposal that California's licensed pest-control advisers be required to file detailed reports justifying each control recommendation, and that each report (emergencies excepted) be submitted to the county agricultural commissioner at least seventy-two hours before intended control action.

These provisions, of course, constitute a clear threat to the freewheeling "advisement" tactics of the pesticide salesmen and to the agri-chemical industry's product-merchandising bonanza. In this connection the provisions have national and international implications. In gathering support for its Sacramento crunch the industry rallied its mob via media channels, exhorting the chemical salesmen to be on hand to protect their jobs and telling growers that the EDF petition was an eco-freak ploy to deprive them of their indispensable chemicals.

At the hearing, industry produced a string of supportive legislators who, emulating Joe McCarthy and Spiro Agnew in their finest hours, rained down a torrent of verbal abuse on the CDF&A panelists. Needless to say, the terrorized CDF&A got the message and canceled three other scheduled hearings on the EDF petition. The agri-chemical industry won the day, while scientific pest control suffered a distinct setback.

Another roadblock to integrated control is that formed by the bloated federal and state pest-control agencies, which dangle the tantalizing pest-eradication carrot before the eyes of politicians and the citizenry. Through this device, these self-serving agencies have gained a hammerlock on massive amounts of public money (tens of millions of dollars per year at the federal level), much of which would be more profitably applied to problem-solving research. Instead, the bureaucracies mostly pound their millions down sterile pest holes in frequently pollutive programs directed against such things as the boll weevil, fire ant, gypsy moth, and tussock moth. As was mentioned earlier, the fire-ant program alone, since its inception, has consumed more than $150 million in public funds, while the ant continues to wave its antennae at its "eradicators." The boll weevil eradication program also continues to burn up its millions of dollars, as do the forest pest-control programs such as the politically inspired $3 million DDT spray program against a Douglas-fir tussock-moth population that had already virtually collapsed from natural causes (see Chapter 8).

About the best that can be said for these programs is that they serve as welfare in the guise of pest control, which perhaps to some addled minds may seem to have social merit in a time of economic recession. But even this dubious benefit is overshadowed by the ecological harm that the programs engender.

Researchers themselves and their institutions form another obstacle to expanded integrated control. Many, if not most, pest-control researchers lack ecological sophistication. They consider their charge to be bug killing and simply do not understand that pest control is essentially an ecological matter: insect population management. Instead, with tunnel vision, they continue to seek simple answers, in which enterprise they are supported by their institutions (e.g., the state agricultural experiment stations and the U. S. Department of Agriculture), which are under constant pressure to develop quick, stop-gap answers to pest problems. Historically, this simplistic approach has been repeatedly manifested in frenetic efforts to exploit innovations: microbial control, pheromones (cue chemicals), autosterilization, pyrethroids, hormones (third-generation insecticides). Each innovation is seized upon as a potential panacea, heavily promoted, supported, and researched, but none ever solves the insect problem. Indeed, they often prove counterproductive by diverting brains, energy, and funds from integrative research.

The grower, too, is a major impediment to integrated control. For example, a number of the most powerful grower associations and crop-producer co-operatives walk in lock step with such potent proponents of the pest-control status quo as the pesticide industry, the food processors, and the pesticide applicators. It is difficult indeed to convince the grower that the people he has been sleeping with for years have in reality been ripping him off.

Even the bankers who finance agri-enterprise create a major roadblock to expanded integrated control. On first impression one would think that the bankers who finance growers would be interested in efficient crop-protection practices, to help assure the security of their loans. At least this is what several leading members of the Association of Applied Insect Ecologists (California's independent pest-control advisers) thought when they arranged a meeting with several San Francisco Bay Area bankers a few years ago. The AAIE people hoped in this meeting to expose the banker specialists in agriculture to the economic advantages of integrated control and the independent pest-control adviser, and thereby develop a favorable climate for financing and expanding their own operations. Along with University of California colleagues Ray F. Smith and Louis Falcon, I was invited to the meeting to provide technical insights into the integrated-control concept.

The meeting was a fiasco. The bankers had no interest whatever in integrated control and the independent pest-control adviser, and gave the impression that they were attending the meeting only as a courtesy to the AAIE members. Finally, in exasperation, one of the AAIE people asked, "But isn't it in your interest to foster efficient pest-control practice among growers that you finance?" He was abruptly answered by the Bank of America man, who stated, "We're not really concerned about the effective use of pesticides; why should we be? We have billions invested in the chemical industry."

The roadblocks to efficient pest control are indeed formidable!

I have painted a grim, perhaps disheartening picture of the potential for rational pest control, and if these were normal times I would expect massive obstacles. But times are not normal; humanity is experiencing unprecedented economic, political, and social upheaval. In this climate of ferment I sense that dramatic changes for the better are in the offing for pest control. The catalysts for change are the deepening food crisis, skyrocketing pesticide costs, and pesticide shortages. We simply cannot continue to play the pest-control game according to the old, simplistic and wasteful, rules. The day of "dirt cheap" DDT and parathion is gone. What's more, the biocides are suddenly hard to come by. The developed nations are going to hoard them and husband their use. Under these circumstances, integrated control may gain increasing acceptance in the First World, but only if there is a loosening of the chemical industry's dominance of the pest-control system. Meanwhile, developing nations short on funds and faced with disease and famine cannot afford the expensive pesticides. Furthermore, those who control these suddenly dear and scarce chemicals will not readily donate them to exchange-poor countries, unless, of course, they get some such organization as the U. S. State Department's Agency for International Development (AID) to foot the bill. Thus, there is a good chance that the international funding and advisory agencies (e.g., FAO, Ford Foundation, World Bank, etc., and perhaps even AID) will be forced to promote integrated control because there is little alternative. Indeed, FAO and AID are now fostering integrated control programs and it is quite possible that under these circumstances the strategy will attain wide-scale implementation more quickly in the Third World than it will among the developed nations.

Chapter 16

TO TURN THE WORM

There are formidable obstacles to change in pest control, and I have tried to show that one of the greatest is the prevailing concept that equates chemical control with pest management. Until all involved elements of society understand that pest management is a complex technology embracing much more than just the poisoning of things, the pro-pesticide bunch will maintain its stranglehold on pest-control strategy. Again I revert to California to illustrate the immense influence of the agri-chemical industry over pest-control practice. Here I have in mind the advertising clout of the agri-chemical industry and its brigades of salesmen, who overwhelmingly dominate the system at the advisement application interface. The chaos that derives from this situation surfaced in the spring of 1975 in the tremendous overtreatment for weevils in California's 1.25 million acres of irrigated alfalfa. In this incident, a major chemical company decided to promote a preventive treatment campaign against the weevils, in direct opposition to University of California entomologists, who recommend insecticide use only when alfalfa weevil larvae approach or attain a certain threatening infestation level (economic threshold). It so happened that alfalfa weevil populations were generally very light in 1975 and most fields were unthreatened by damaging infestations. Nevertheless, vast acreages were needlessly treated, because many growers, hustled by salesmen or swayed by advertising, "bought" the chemical company's preventive treatment program. The result: needless expense, a waste of valuable chemicals, and extensive, pointless environmental pollution. What I am stressing here is that all of this waste and pollution occurred because a chemical company had greater influence over pest-control decision-making than did university insect-control specialists and independent pest-management advisers. More recently, as noted in this book's Introduction, this same company promoted a similar preventive-type chemical control program in another crop. Here again it acted counter to the program advocated by university research and extension personnel, a program that was arduously developed to bail the growers out of a serious secondary-pest problem.

After witnessing this type of pest-control chaos for more than a quarter century, it is abundantly clear to me that the elimination of the pesticide salesman from pest-control advisement is absolutely necessary if we are to develop a better pest-control system. It makes no more sense in the complex matter of pest control for salesmen to diagnose crop ailments, prescribe chemotherapy, and then peddle their own pills than it would be for pharmaceutical salesmen to function similarly in human or veterinary medicine. In fact, the medical profession, which recognized this problem quite early in its evolution, does not allow the pharmaceutical industry to dominate diagnosis and prescription. Although M.D.'s and D.V.M.'s are human and have their shortcomings, I for one greatly prefer having them, rather than drug salesmen, diagnose and prescribe treatment for my ailments and those of my dog.

The Pesticide User

Whether it's agri-macho, techno-fascination, or simply the hunter-killer instinct, there is something in us that compels us to kill bugs. And this compulsion is very much a factor in the massive overuse of pesticides. Of course the agri-chemical industry is aware of this quirk and gears its advertisements to titillate our lust to kill. The ads read:

Morestan® murders Mites.

Hit'em with Lannate®.

The shotgun and the rifle. Diazinon® the shotgun,
Galecron® the rifle.

Move'm out with Dylox®.

But the killer lust is ours, and it is we who bear direct responsibility for the pesticide overuse that it engenders.

This point was recently brought up in conversation by Dr. Lynell Tanigoshi, a researcher in integrated control of apple pests at Washington State University's Wenatchee experiment station. Tanigoshi was telling me about the benefits of Washington's highly successful spider-mite integrated-control program in apple (see Chapter 15). He said that refinements have made the program so effective that where growers do the proper things, they do not have to spray for spider mites at all. This is a remarkable change in a pest problem that once threatened Washington's entire apple industry. But then Tanigoshi sadly shook his head and said, "It's crazy, Van; you just can't get some of the growers to follow the integrated program, after ten years of success. It seems like it's in their blood to crank up their rigs and go out and spray the groves. And when they do this they foul things up. I don't understand it. They completely forget that just a few years ago the apple orchards of central Washington were burning up with spider-mite infestations created by excesses in spraying practices."

Dr. Robert Luck, a young researcher at the University of California, Riverside, tells a similar tale about California citrus growers: "You know, Van, if they use the information that we have developed for them, those guys probably only need to spray for red scale once every two to three years. But they don't use the information; they just keep on spraying every year. Because of the economic constraints of citrus production that make the grower very apprehensive about insects, plus the lack of support for basic ecological assessment of the citrus-pest situation, I am discouraged to the point where I am seriously thinking about giving up my research in that crop. There are more rewarding things to do."

Mind you, the citrus agri-macho spends seventy-five dollars or more per acre each time the spray rig rumbles through his grove. Whatever it is, the inner urge to spray even overwhelms his sense of good economic management!

I once felt that if a man wanted to squander his money on needless pesticide spraying and simultaneously risk turning on real problems, it was his right to do so. After all, this is a free country and it's one's prerogative to do foolish things on one's own property. But, in giving careful thought to the matter, I have arrived at the conviction that with pesticides, foolish use is not an inalienable right.

There are multiple reasons for reaching this conclusion, but mostly it derives from the fact that almost inevitably pesticide spraying has impacts that extend beyond the point of application. In other words, the fool's act is not restricted to his own property but extends beyond it to hurt other persons and the environment.

For example:

1. Where crops or other large tracts such as forests are treated there is always a high risk of pesticide drift to adjacent areas and in some cases movement to remote places. Excessive pesticide use inexcusably increases this hazard.

2. Excessive spraying increases genetic selection for pest resistance to pesticides. This has already had tragic manifestation where profligate agricultural use of pesticides has contributed to resistance in disease-transmitting mosquitoes. And in the agricultural sector itself, excessive pesticide use has unquestionably contributed to the acceleration of resistance in a number of crop pests. In both cases society suffers for the sins of a few.

3. Excessive use of pesticides increases adverse societal and environmental impacts even where the chemicals remain at the site of application. For example, most people injured by pesticides are persons who handle the materials, or work where they are applied. Thus if the chemicals are being used in excess of actual need, all poisonings resulting therefrom represent a needless social impact. From the environmental standpoint, losses to pollinators and impacts on widely ranging wildlife that result from excess pesticide use are also inexcusable.

4. The needless use of pesticides represents a squandering of materials derived from non-renewable resources, a waste that is intolerable on moral grounds.

In my opinion, the grower (farmer) in particular has a responsibility to society to clean up his pesticide act. This thought will undoubtedly raise the hackles of the typical farmer, for he is a fiercely proud and independent person who resents others' meddling in his affairs. But as proud and independent a free enterpriser as he may be, the grower owes something to the bird and bunny lovers, clean-air addicts, pure-food freaks, organic-gardening advocates, consumerists, environmentalists, eco-radicals, political activists, farm workers, and all the others who are concerned about the pesticide overload, because as tax payers they contribute inordinately to his ability to operate and survive under our socioeconomic system. It is they in their collective millions who provide the bulk of the dollars for government crop price supports, the vast irrigation and soil conservation systems, the land-grant universities, and USDA research, which are so vital to American farmers' economic health, and it is they who endure the elevated food costs that result from the monopolistic provisions of the federal marketing orders. Efficient and safe use of pesticides is a small concession for the bulk of us to seek from a segment of society that benefits so substantially from our economic sacrifices.

I would hope that I have made the point that the individual pesticide user, be he or she a home gardener, cotton farmer, mosquito abater, or forest pest-control specialist, should somehow be constrained in the freedom to use pesticides. In the field of medicine, only the physician is allowed to diagnose the need for and to prescribe the use of drugs and medicines. And these are substances that, if misused, by and large hurt only individuals. But, in pest control, a person wasting his money and energy on needless spraying loads the environment with chemicals that can hurt or kill other persons and other creatures. What is so disturbing about this is that the pesticide profligate violates our molecular privacy with impunity. In other words, the general populace is exposed to and often absorbs molecules that, whether benign or potentially harmful, are recklessly dumped into the environment by persons who cannot be held to account. Every individual has a right to maximum molecular privacy, and it is society's responsibility to guarantee that right. But, never mind this right; so long as powerful economic and political forces favor the status quo, we will be under extreme pressures to hold to our mad pattern of pesticide use.

The Land-grant Universities

Clearly, we must get both the pesticide peddler out of diagnosis and therapy-prophylaxis and the compulsive pesticide user-client off his chemical fix, and replace this system with a system in the mold of medicine or veterinary medicine. But I reiterate that this will probably occur only when society realizes that pest control is something more complex and demanding than mere bug killing. Social realization implies an educational campaign, and as such, is a major responsibility of the land-grant universities.

But here again there is a formidable obstacle, for as I mentioned earlier, much of the land-grant university apparatus is solidly aligned with the pesticide bunch (remember CAST and CEFAFP!). Furthermore, even where there is sentiment to abandon the chemical strategy, the pro-pesticide interests, as already noted, often have the power to force adherence to the status quo. So the land-grant university and its programs and policies represent another area where change is necessary if a better pest-control system is to evolve.

By and large, the aggie colleges and their associated experiment stations and extension services are social anachronisms that view their mission as one narrowly oriented to crop production and agri-business and hardly concerned with broader societal interests. What else explains their virtual neglect of the concerns of the farm worker, the consumer, the urban homemaker, and the environmentalist? This narrowness is perhaps explainable in largely agricultural states, where the universities are dominated by farming interests. It is difficult to envisage the evolution of society-oriented programs in these institutions. But in highly urbanized states such as California, New York, New Jersey, Pennsylvania, Illinois, Michigan, Ohio, Florida, and the like, there is every reason to expect that the aggie colleges can and should be upgraded both in the scope and nature of their programs and in the quality of their research and teaching, so as to address themselves to the related problems of society at large. And it is in such places that (it is to be hoped) a better pest-management system will evolve. Good pest management is in fact in the interest of all segments of society, for efficient food and fiber production, the safeguarding of human health and comfort, and the maintenance of a quality environment are societal goals, and pest management should be researched with this in mind.

What is needed, then, in the more promising institutions is not only an upgrading of the quality of ag-college pest-control research, teaching, and extension staffs, and students, but also a broadening of orientation and mission. In this process there must be a ruthless culling of low-quality and obsolete staff personnel, and the recruitment of bright, dynamic, technically competent, sociologically and ecologically aware young people to fill the vacancies. Furthermore, where tenured drones hang on, they should be isolated and kept away from policy and decision-making roles. This is a tough way to treat good ol' Joe down the hallway, but it must be done, for he is an impediment to progress and, as such, an intolerable social liability.

Here I speak from firsthand experience, for I have spent my entire professional career as a land-grant university researcher and teacher and have watched in dismay the fossils (both mental and chronological) throw up their roadblocks to pest-management modernization. Low-quality faculties conduct pedestrian research and do low-grade teaching. Furthermore, at the graduate training level they turn out a pedestrian product that recycles back into the system to perpetuate mediocrity. The generally low quality of entomology faculties is reflected in the roster of entomologists in the National Academy of Sciences, which lists only a couple of land-grant university types among the more than twenty entomologists in the Academy. As one perceptive young colleague puts it, "The definition of the word entomologist is 'dumb biologist'!" Unfortunately, as I ponder the entomology faculties with which I am familiar, I must largely agree with my brash young friend.

But it doesn't have to be this way. Entomology and specifically pest-insect management can and should be an exciting and intellectually challenging profession. The same holds true for other disciplines involved in pest control, such as plant pathology (plant disease control), nematology (control of plant parasitic worms), and economic botany (weed control). There is a tradition for the best brains in science to go into disciplines such as physics, chemistry, mathematics, physiology, biochemistry, and genetics, to the neglect of applied biology. But now that we have discovered the double helix, probed the nearby planets, developed the SST, conquered polio, and invented television, the overwhelming challenge that confronts us, aside from human population control, is the need to feed, house, clothe, and protect from pestilence our burgeoning billions. And here one of the greatest challenges is how to keep pests from robbing us of almost half the food and fiber that we grow.[111] What I am driving at is that pest control badly needs an infusion of superior scientific brain power, for until we bring this intellectual weaponry to bear against our "clever," resource-raiding antagonists, the latter are going to steal us blind.

I don't really know how reform will occur. But I do detect a developing interest among very bright young scholars in matters concerning resource development, production, and protection, and this is a ray of hope. In fact, at Berkeley these "strangers" are beginning to show up in entomology, and I suspect that this is true elsewhere across the land. I would hope that administrators and faculty scattered through the land-grant university system sense this trend, read its significance, and at least in some of the institutions, undertake the upgrading of their pest-control research and teaching programs. This, of course, will require internal adjustments within the universities, especially in the administrator mentality, but it will also depend upon a sustained awareness by the political prime movers as well as the policymakers of both governmental and private granting agencies, which must help underwrite the new training and research programs. Here I hasten to point out that I do not advocate the pork-barreling of grant funds to every cow college that puts out its hand but, instead, a careful allocation of support funds to those institutions with quality programs that will serve the best over-all interests of society. And to be fully objective, I envisage that such institutions need not necessarily be land-grant universities.

But even assuming that we do develop a system that produces a quality product, there will still be formidable obstacles to getting that product into the mainstream of pest management. It is my impression that the majority of the agricultural experiment stations and the U. S. Department of Agriculture are not ready to accept a new breed of dynamic, questioning researchers and teachers, and until they are, the modernization of pest management will be greatly encumbered. For example, I've known a number of bright young Ph.D.'s who couldn't or wouldn't adjust to the bureaucratic straitjacket of the ag colleges and the USDA, which together constitute the greatest job pool for research entomologists. These young people simply refused to knuckle under to aggie-college or USDA conformity. In other words, under the status quo there is a tendency for quality scientists to shun the USDA and agricultural-college bureaucracies and drift off into more intellectually rewarding areas. Somehow they must be induced to remain in the pest-control arena. To accomplish this, I would suggest that until there is a substantial philosophical change, particularly in the aggie colleges, a framework of medium-term fellowships be built around research programs in the quality institutions so that the best of the young people can be kept in pest management and then fed into the system as it evolves away from the prevailing pattern. This is a matter that should be given serious consideration by the Congress, the state legislatures, and agencies such as the National Science Foundation, EPA, the National Institutes of Health, and private agencies such as the Ford and Rockefeller foundations. It is particularly important that the latter, which have helped foster the Green Revolution, should assume a substantial responsibility for modernizing plant-protection philosophy and practice in the developing nations, for otherwise the pesticide mafia may well move with disastrous effect into the technological strip mine created by the Green Revolution. A precursor of the kind of chaos that can develop in conjunction with agri-chemical company dabbling in the Green Revolution has already occurred in Indonesia.[112]

The United States Department of Agriculture

The U. S. Department of Agriculture is a particularly serious impediment to the modernization of pest management. Its pest-control policies under its research branch, the Science and Education Administration (SEA), and its pest-control operations branch, Animal and Plant Health Inspection Service (APHIS), approach a hopeless muddle. USDA pest research and control agencies are so badly riddled with and driven by external and internal politics, and encumbered by bureaucratic inertia, that they seem incapable of running coherent and meaningful programs. For example, for at least a decade enormous amounts of manpower, brain power, money, and facilities have been squandered by SEA on backup research for APHIS-conducted, politically inspired pest-eradication and area-control programs, while other research areas have been neglected even to the detriment of staff morale, not to mention the public welfare. Bright young entomologists typically turn to USDA employment only as a last, desperate resort, if at all. And once within the edifice, they too often slip into pedestrian ways after a few traumatic confrontations with the ponderous bureaucracy. The brave ones eventually get out.

SEA, in particular, badly needs a thorough house cleaning, but this seems virtually impossible. Powerful congressmen have too much control over the agency to permit it to be modernized and molded for efficiency. Traditionally, the politicians who dominate SEA are the farm-state dinosaurs who use the agency for pork-barreling and "welfare" programs back home (e.g., a massive insect physiology laboratory in some prairie-state boondock; a multimillion-dollar pest-eradication program somewhere in the "hush puppy" belt). The horrors of life in the politician-terrorized USDA were brought home to me on the occasion of a visit to the Delta States Agricultural Research Center, in Stoneville, Mississippi, where I had gone to present a seminar. I can only describe the experience as saddening, for in all my adult life, during which I spent a year in the war zone of the South Pacific as one of a shipload of frightened souls under recurring showers of kamikaze aircraft, and later a half dozen years in countries suffering various forms of political tyranny and socioeconomic misery, I have rarely encountered a group so demoralized and embittered as the entomologists at Stoneville. Some of those entomologists were personal friends of long standing.

The Delta States Agricultural Research Center is a monument to the ego and political clout of its principal congressional protagonist, who reigned for years as one of the most powerful and feared men in the U. S. Congress. The facility, a high-rise concrete monolith, bulks up out of the black dirt flatlands of the Mississippi Delta like a displaced Ayers Rock. And like Ayers Rock, there is no real reason for its being there, because Stoneville, Mississippi, is a rural outback remote from any major academic institution, cultural center, or communication network, and as such the worst possible place for creative, dynamic research and the cultural needs of high-quality research scientists.

The Delta States Agricultural Research Center enjoys its improbable siting at Stoneville because the congressman wanted it there; it is his home turf. In other words, the concrete monolith rises out of the Mississippi Delta as a classic monument to political pork-barreling. The feelings, working atmosphere, and psychological well-being of the scientists assigned to the facility are entirely secondary to this reality.

Stoneville's inherent shortcomings as a working/living atmosphere are bad enough, but what makes the place even more depressing is the embittered mood of many of the staff who were dragooned or coerced by craven USDA administrators to accept assignment there. The problem that the administrators faced was that, upon completion of the center, they had a giant facility on hand, no money with which to staff it, and a powerful, truculent congressman on their backs, antsy to get his prized colossus on line. There was only one thing to do: just go out and corral bodies from the existing USDA entomology roster and dump them, kicking, scratching, and screaming, into Stoneville. And so this is what was done. People were simply plucked out of USDA laboratories in such places as Montana, Arizona, California, South Carolina, New England, etc., and like it or not, plugged into the Delta States Agricultural Research Center. These were career professionals, some with many years of experience in the special problems of their areas and with deep social and cultural roots as well as economic interests (e.g., their houses, other properties, their children's schooling) in the areas of their long-time residence. In the cruelest possible way, they were uprooted and deposited into a nightmare world where, among other things, they were social outcasts, fitting neither with the resident white aristocracy, the lower-class whites, nor the black community. I had long known some of these "displaced" persons and was sickened by their personal tragedies. Among my personal acquaintances, one man's wife simply refused to accompany him to Stoneville, and divorce ensued; another wife suffered a nervous breakdown during the ordeal. Several of the researchers, previously productive scientists, told me that rather than spend the rest of their useful years in Stoneville, they would sweat out the necessary number of years and take early retirement, at considerable economic sacrifice.

Shattered lives!

But the congressman doesn't care, he has his monument fully stocked with GS-11's, 12's and 13's; and the administrators don't care, they have the congressman off their backs.

At Stoneville I gave my lecture, went through the mechanics of visiting laboratories and consulting with staff members, and then gratefully boarded my plane and got the hell out of there! But I am still contaminated by the place, because I cannot erase the memory of proud, competent scientists in their middle years, frustrated and defeated, not unlike men serving time for crimes they did not commit.

Stoneville is perhaps an extreme example of the evil of the politician/administrator interface in USDA, but it symptomizes a widely systemic malaise, for USDA pest-control administrators know the rules of the game and play them to perfection. In SEA and APHIS, pork-barreling and welfare masquerading as pest-control programs prevail, while science, scientists, and society are shortchanged.

The Carter administration, through Secretary of Agriculture Bob Bergland, is attempting substantial reform of USDA pest-control policy, but these noble intentions are confronted by an entrenched bureaucracy that is most comfortable with the good old ways and prepared to dig in and hold the line until the right kind of folks regain control of the White House.

It is thus doubtful that major change will occur in USDA pest-control policy within the near future.

Despite this, it is my contention that what the country badly needs at the federal level is some sort of high-quality national pest-management research institute on the model perhaps of the National Institutes of Health. Within such a framework, federal pest-management research could operate at a high scientific level, free of the direct political pressures that now overwhelm SEA. Under such a system, federal research could be directed to basic aspects of major problems of national and societal importance, while the state experiment stations could address themselves largely to the more parochial problems of their areas. The federal institute could also provide research grants to various institutions, much as do the National Institutes of Health. There would be several advantages to this system. For one thing, it would operate with greater efficiency than the prevailing system, in which SEA and land-grant university research is often duplicative and in some cases competitive. Under such a new system, augmentative and co-operative research between the federal and state agencies would evolve much in the pattern of the research programs of the National Institutes of Health and the various medical schools. Quality researchers would be attracted to the federal research institute, and they would work in a stimulating and rewarding atmosphere. Finally, federal monies now often squandered on pork-barrel, "welfare," and duplicative programs would be channeled to the state experiment stations or consortia of stations to augment research efforts on serious or acute local or regional problems. In England, for example, there is a National Agricultural Research Council under the National

Science Research Council, which awards research monies to various private and public institutions. The Agricultural Council is composed of highly respected and qualified persons who critically review research proposals and award funds on the basis of merit. It also serves to buffer the institution and researchers from direct political interference. Under this system, even such unlikely institutions as Cambridge University and the University of London do agricultural research, and it is very good stuff, too!

The suggested national pest-management research institute plan envisages a severe curtailment of the traditional SEA pest-control research activity. This cutback, of course, would be anathema to farm-state politicians and SEA bureaucrats, and for that matter, to the pesticide mafia, which largely manipulates these people. It is inevitable that these groups would fiercely oppose such an evolution and, to be quite frank, I am sure that in the prevailing climate the odds are that they would overwhelm it. In making this proposal I have had two objectives: first, to point out that in my mind there is indeed a better way to go about pest-control research at the federal level and, second, to stress again the enormous obstacles that confront the development of a rational pest-control strategy in the United States.

The Pest-control Professionals

We scientists involved in pest-management research must also undertake some deep soul searching if a better bug-control system is to evolve. In particular, we must divest ourselves of the corruptive influence of the agri-chemical industry and its allies and our distorted commitment to chemical control. Although I will use the entomology profession to illustrate this point, I hasten to note that other professions, such as plant pathology, weed science, forestry, and agronomy, are perhaps even more in need of self-catharsis.

Lamont Cole alluded to the prime movers in the Entomological Society of America (ESA) as a coalition of "chemists, toxicologists, and others primarily concerned with the destruction of insects." This is probably overstating things a bit, but with the insecticide as the dominant tool of destruction it is inevitable that the agri-chemical industry should have substantial influence over the entomologists and their Society. A peek into the Entomological Society's operations gives some insight into this influence. For example, at ESA's annual meeting among the highest honors bestowed on "deserving" members are the CIBA-GEIGY (Founder's) and the Velsicol (Bussart) awards, which entail cash prizes and, in one case, a company-supported overseas trip. In the event that the names just cited are unfamiliar, CIBA-GEIGY and Velsicol are major agri-chemical producers.

Then there are the Society's Sustaining Associates, a substantial roster of the nations's agri-chemical companies who aid the Society financially by paying inflated annual dues (one hundred dollars or more per membership), helping pay the freight for "distinguished" speakers, and advertising in the Society's journals. As their reward for this patronage, the Sustaining Associates are conspicuously listed in the Society's bulletin and its annual meeting program. From this, we rank and file members get the message that these folks are *family*.

At the Society meetings, certain of the companies operate hospitality suites to which lucky members are invited to savor a variety of prime-quality booze (you will remember from an earlier chapter that I do not have a hang-up over booze) and indulge in conversation and camaraderie with other select entomologists and company brass. And if one is especially deserving, he is invited at company expense to join an elite group of colleagues and hosts for dinner in some elegant restaurant.

Aside from its intrasocietal influences, the agri-chemical

industry also hands out its largesse to "deserving" entomologists on an individual or group basis. For example, it is not uncommon for "meritorious" researchers or farm advisers to be given full-expense packages to attend meetings. What's more, especially "deserving" entomologists may be awarded overseas or even 'round-the-world trips, and for the less lucky there are paid-for fishing or hunting holidays. For example, in California one chemical company routinely pays the bills for "deserving" researchers and farm advisers to go on fishing holidays to La Paz, in Baja California, Mexico. At one southern university, the Entomology Department's outstanding annual social event is its two-day squirrel hunt, during which a chemical company provides the good ol' boys with free booze and vittles.

Much to my amazement, when I have discussed these "little" rewards with the lucky recipients, they deny that the favors constitute a form of corruption. "Gollee, Van, some little ol' gesture of friendship or appreciation cain't corrupt me." These folks sound just like those duck-hunting admirals and generals of recent fame!

Another way in which the agri-chemical industry helps "deserving" entomologists and their institutions is through mini-research grants. Almost every one of the country's entomology departments has a bagful of such grants, and some virtually depend on this type of support to carry on their programs. These grants are usually specifically earmarked for research on the donor's proprietary materials. And where federal registration or university recommendation of the pesticide is the objective (as it usually is), the companies expect good treatment from the grantees in return for past, current, and future generosity. However, on occasion the grantee doesn't perform as expected and there is hell to pay. I recall two cases of company reaction to grantee "malfeasance." The first involved my colleague Dr. Louis Falcon, who had a small grant from a pesticide company to assess one of its microbial products. At a certain point in time the company attempted to register the material for use against a cotton pest in California, and in connection with its petition submitted performance data derived from studies made in another state. The California State Department of Food and Agriculture contacted Dr. Falcon and asked his opinion regarding the adequacy of the data. Falcon responded that in his opinion the data were grossly inadequate, and on this basis registration was denied. I was having lunch with Falcon on the day that the pesticide company "hit squad" ran him down and suggested a little talk. The scene resembled something out of a James Cagney gangster movie of the 1930s. The company boys told me to get lost and then escorted Falcon to a corner booth, surrounded him, and proceeded to verbally abuse him for his transgression. They didn't physically beat him, but they "leaned" on him very hard vocally and then told him that they were going to take away his grant. Lou Falcon, a gentle, sensitive, and idealistic person, was deeply affected by the muscling he had endured. When the torpedoes departed and he rejoined me, he was visibly shaken. He told me then that he would never again accept a chemical-company grant.

The second incident involved Dr. Charles Schaeffer, another university colleague. Dr. Schaeffer had received a mini-grant from a chemical company to conduct studies on one of the company's "hot" experimental materials then being considered for federal registration and labeling. In the course of his study, Schaeffer, an experienced, competent, and conscientious researcher, found that the pesticide had a serious environmental flaw, which he called to company attention. It turned out that the company was already aware of the problem and had been covering it up. Company management, furious about Schaeffer's discovery and his indignation over their cover-up, reacted by taking away his grant.

Louis Falcon and Charles Schaeffer played it straight

with their mini-grantors and were punished. This raises the question: just how many grantees are as honest as Falcon and Schaeffer, and how many, to safeguard their grants, play the game by company rules?

Additional evil in the chemical company mini-grant system is, first, that it channels too much research down the chemical line; second, that the relatively few dollars proffered tie up inordinate amounts of the grantee institution's resources; and third, that for "pennies" the grantors buy university endorsement of proprietary products. Very few deans, directors, chairmen, or researchers acknowledge this rip-off for what it is.

The chemical industry mini-grant program has its counterpart in the system of grants and gifts offered by the agri-commodity groups (e.g., the grain growers, pepper packers, potato producers), which buy up researchers and their institutional support with small gifts and grants mostly earmarked to benefit the particular commodity group. I am aware of one entomologist who has devoted the bulk of his quarter-century research career (and the current fifty thousand dollars per year in public funds it represents) to a commodity group that up until last year awarded him an annual mini-grant of less than ten thousand dollars a year. For the growers, that's a pretty good return on a small investment! Unfortunately, support of this sort is being increasingly sought by researchers, because, as is the case in California, traditional funding sources are evaporating. For example, I have been forced to go this route with the alfalfa weevil problem, because there is absolutely no hope of obtaining emergency funding from the state legislature. But I insist on setting the specifications on my research programs, and if these stipulations are not met, I will refuse the grant. This is the understanding that I have with the California Cattlemen's Association. In the case of a cotton research grant I once held, pressures mounted for me to do things in specified ways. I refused to do this and, much to the dismay of university administration, rejected the grant, which was of several years' standing. When I was at the University of California, Riverside, a number of years ago, the citrus industry, through university administration (the college dean), tried to force me to abandon much of my existing research program to address a problem in citrus. I refused to do this and got in a terrible hassle with the dean and my department chairman. I won't go into the details of what they did to me, but I paid a price. Not all researchers are willing to do this, and as a result, too many are ultimately "bought up" by the commodity groups, to the detriment of society's interests.

There are, of course, ways to minimize or even eliminate the evils of the agri-chemical industry and commodity-group mini-grants. For one thing, the grantee institutions can establish basic conditions to be met before such grants are accepted. Better yet, where an institution has a pest-management master plan, all incoming grants, regardless of source, could be pooled in a central fund to be allocated to the various research areas as the over-all needs of the program dictate. This, of course, would result in a drying up of many of the agri-business grants and gifts, but it would in turn get a major parasite off the institution's (and society's) back as well as maximize the impact of "sincere" gift and grant monies.

The real answer to the problem of "earmarked" gifts and grants is institutional rejection of such monies. But this is asking a lot of the administrator mentality at a time when budgets are tight and money is where one finds it. In the final analysis the answer lies with society, in that *it* has to decide whether it wants a quality pest-management program and, if so, whether it is willing to pay the price. Meanwhile, those of us in research and extension, as well as society in general, must recognize the hazards of the prevailing system and bend every effort to prevent its becoming a one-way bonanza. All that I can do at this stage of the game is point to and shout about the evils of the mini-grant system.

Insect Eradication and Area Control

It is a sad reality that some of the greatest pesticide abuses and/or wastage of public pest-control funds occur in the insect-eradication and area-control programs so deeply cherished by the federal and state pest-control bureaucracies. These programs are also the most difficult to discredit, because, as touted by their proponents, they seem to have so much going for them. For example, the glittering prospect of cotton boll weevil eradication has enormous appeal to growers and politicians and even environmentalists. Why, as the USDA pest eradicators tell it, the expenditure of just a few hundred million, or perhaps even a billion, dollars may forever eliminate a devastating pest that has already cost society tens of billions of dollars and contributed to massive pesticidal pollution. Mind you, these are the same USDA pest controllers who showed their concern for the environment by siding with the agri-chemical industry in its battles to prevent EPA from banning DDT, aldrin-dieldrin, chlordane, and heptachlor. And they are the same people who have spent more than $150 million "eradicating" the fire ant by raining down ton upon ton of two hazardous insecticides, dieldrin and mirex (the latter a first cousin of Kepone®), over millions of acres of the Southeast. Beware, Mr. Taxpayer!

There is nothing wrong with the eradication concept or, for that matter, area-wide pest control, but each of these tactics has its time and place for maximum applicability and efficacy. Successful eradication programs have been mounted against populations of such pests as the khapra beetle, the Mediterranean and oriental fruit flies, the Mexican bean beetle, the mosquito *Anopheles gambiae,* and at places the screwworm. But these populations all had exploitable weaknesses. In other words, they were relative pushovers; but unfortunately this is not true of most pest insects. What has gone wrong with the eradication concept is that it has fallen into the hands of persons who are more skilled at program promotion and empire building than they are at the business of scientific pest control. With these people the program becomes more important than its objective, and so it typically goes on and on and on, gobbling up funds and polluting the environment but rarely realizing the original goal, of pest eradication.

Even the screwworm program, the showcase "success" of the eradicators, is beginning to assume this pattern. In this case, release of sterile male screwworm flies had for years maintained the pest at very low levels, saving the Texas cattle industry over $100 million annually at a cost of about $10 million in taxpayer dollars. But then, suddenly, the control broke down, and there was a resurgence of screwworms to some of the highest levels ever recorded. The cause of the breakdown appears to have lain in the mass-reared flies, which after years of soft living had become lazy, fickle in their sex habits, and somewhat blind.[113] The USDA was unaware of this situation and reacted to its exposure by contending that the failure lay not in their screwed-up screwworm flies but, rather, in the recurrent invasion of Texas by flies boiling up out of Mexico. It then made a bizarre proposal: cut the flies off at the Isthmus of Tehuantepec (extreme southern Mexico) and then march north from the isthmus and south from Texas and sterilize them out of existence in Mexico.[114] The asking price: additional millions to come out of the U.S. taxpayer's pocket. This amount added to the money already being plugged into the program brings the total cost of screwworm eradication to about $20 million a year. But how much longer should the public be asked to pour its millions of dollars into the elimination of an insect that was pronounced eradicated a decade ago? When one ponders the enormous task of dropping sterile male flies over the vast, mountainous land mass of Mexico, where the screwworm breeds the year around, it would seem, as far as the USDA is concerned, forever! Incidents such as this leave one with the impression that insect eradication is very much akin to that elusive "light at the end of the tunnel" that led us down such a tragic and costly pathway in Vietnam.

From what has just been outlined, it should be evident that since insect eradication is a concept on the verge of going wild, it is in urgent need of containment. But such discipline will be extremely difficult to enforce in the existing political climate, influenced by powerful legislators who encourage and support the federal and state pest-control bureaucracies in their eradication and area-control activities. And why shouldn't they? The programs are wonderful sources of vote-winning propaganda, not to mention political pork and welfare funds. So, again, I will suggest a remedy that even if it has merit, has very little prospect of ever receiving serious consideration.

But here goes anyway!

To my mind the critical step needed to bring balance to the eradication/area-control situation lies in the establishment at both the federal and state levels of some kind of watchdog mechanism (board or commission) to oversee the conception and operation of the programs. Such a body would have, as its ultimate prerogative, veto power over program initiation and program continuation. As matters now stand, the programs are almost always internally conceived, with political backing and then internally "policed." And under this arrangement, policing means essentially program perpetuation and expansion.

An effective watchdog commission, then, would have to be appointed in such a way that there would be little or no input into the selection process by the affected agencies. Perhaps this could be done through the medium of the National Academy of Sciences. It would also be highly desirable that such a commission be composed of members representing a spectrum of societal interests; e.g., agriculture, consumerism, environmentalism, and public health. In other words, the board(s) should be composed of technically competent, objectively motivated members free of pressure and influence by the affected agencies or special-interest groups.

I fully realize that what I have just suggested borders on fantasy, not because it lacks merit but because there are so many forces in society that would work to prevent its being realized. I have simply thrown out for consideration the suggestion that there is a way to adjust a serious societal dilemma. In the final analysis, it is up to society to determine whether or not it has the will and capacity to modify an excessively wasteful and pollutive practice.

Safe, Selective Insecticides

The final point I wish to touch upon in this discussion of pest-control reform is perhaps the most important of all. Here I refer to the can of worms entailed in the development, registration, production, and utilization of pesticides. But first I wish to stress that under prevailing conditions *we need pesticides*. Pest control is one of the overwhelming realities of our time; pests gobble nearly half of what we grow, and they pose an immense threat to human health. In other words, pest management is one of the paramount concerns of humanity, and since we are currently locked into a chemical control strategy, we simply cannot drop this strategy overnight without suffering unimaginably. What's more, our crops and livestock are largely genetic artifacts that have to be coddled if they are to perform effectively or even survive. We also tend to grow our crops in monoculture, which sets things up for devastating pest outbreaks which, when they occur, can, under prevailing technology, be beaten back only by chemical sprays. Then too, our own finicky requirements for cosmetic, pest-free produce have led to the imposition of impossibly low pest-tolerance standards, which force the use of pesticides. Finally, as the human population burgeons and thus creates increasing need to protect our food and fiber sources from competitors, and ourselves from nuisance pests and disease transmitters, the pressures to spray are intensifying. The basic argument, then, is not against pesticides per se or chemical control as a tactic but, instead, against chemical control *as our single-component pest-management strategy and the biocide as its operational tool.* To repeat a theme of this book, it is the chemical control *strategy* which has gotten us into serious trouble with the insects, and unless we abandon this strategy things will only get worse. But along with this strategic change it is also vital that society insist upon the establishment of standards that eliminate "biocides" as our chemical tools, and require, instead, safe, selective, and ecologically tenable pesticides.

The chemical industry and agri-business will howl that industry cannot and will not produce such materials because they are too costly and too limited in their marketability. This is hogwash (or is it sheep-dip?), and there are very clear and logical reasons for my saying so. Most importantly, there is a multibillion-dollar market out there, and as long as that kind of gravy exists, free enterprise will manage to get at it even if it takes safe, selective pesticides to do so. The problem today is that there are too many companies with too many products battling for the swag; fourteen hundred pesticides and thirty thousand labels. What a joke! This forces the chemical companies into a merchandising dogfight and into continuously seeking another DDT or parathion; that is, a low-cost biocide designed more to capture markets than to fit into scientifically conceived, integrated pest-management systems.

But if society, via legislation, demands safe and selective pesticides, the chemical industry will adjust to that reality and provide the materials, simply because a billion-dollar market awaits such products. A number of events can be expected in conjunction with a transition to safe, selective pesticides. For one thing, many companies lacking the financial and technical resources to produce sophisticated materials will be forced out of the insecticide business. But this is hardly lamentable, since there are too many pesticide producers as it is. Increased pesticide cost is another sure eventuality, and despite the inevitable wailing of the hand-wringers, this is desirable, for it will force us to use pesticides more judiciously. Thus, despite their higher price, chances are that the safe, selective pesticides will actually bring about a reduction in over-all pest-control costs because of more judicious use. But even more important, they will fit neatly into integrated pest-management systems, which, experience has already shown, employ insecticides more effectively, in lesser amounts, and at less cost than do conventional chemical control programs. An important consideration here is that the safe, selective materials would not engender the costly and disruptive pesticide treadmills that characterize use of the biocides, and thus would not create the severe ecological and social impacts that constitute the tremendous hidden costs of conventional chemical control.

Society's Choices

The "solutions" I have discussed in this chapter are perhaps impractical or even naïve in light of the prevailing economic and political climate of contemporary America. But they represent a conscientious attempt to offer constructive alternatives to a shocking state of affairs in an important area of technology. I can only hope that I have presented these alternatives wisely and well and that my ideas will have some influence on public thinking and action. Society gets what it wants or what it deserves, and too often, in either case, it and the environment are shortchanged. Therefore, as we struggle through the late-twentieth century with that monster of our own creation, technology, it is increasingly important that we assess its applications and implications wisely. And in doing this it is especially critical that we have the patience to listen to all reasonable voices and develop a broad information base to guide us in our decisions. A technological innovation once applied may have irreversible impact, and so it is the enormous responsibility of humanity to weigh technological decisions with utmost care, for the Earth can survive just so many adverse techno-impacts.

EPILOGUE

The Voices of Nature

We as a species have many vices, and among these, corruption is one of the deadliest. The previous pages have detailed how corruption has penetrated just one human enterprise, pest control, and rendered it a near shambles. The taint is everywhere: among politicians, industrialists, merchandizers, food processors, government and univerity administrators, government and university researchers and extension specialists, federal, state, and local pest-control agencies, pest-control advisers, pest-control applicators, growers' organizations, and elements of the media.

This malaise is global in extent. Everywhere I have gone in the developing countries, when I have asked the question of respected colleagues, the answer is always the same: "Yes, corruption is commonplace."

This is the state of affairs in just one small area of human technological endeavor. But it symptomizes what must be going on in every branch of applied technology. This is why corruption is so deadly. It cripples our ability to responsibly assess what technology is doing to our planet and thereby to our own survival as a species.

Nature is emitting signals warning us that under the existing format the future is ominous. She is saying that we cannot continue our attempts to ruthlessly dominate her and that if we persist disaster is in the offing. She has many voices, and one of the clearest is that of the insects. The insects have already told us that we cannot overwhelm them and that there has been a price to pay in trying. But Nature has other voices, and if we listen carefully we can hear these additional warnings too: the voice of the trees in the crashing of a forest before an assault of axes, and the later rumble of a mud slide as a cloudburst sweeps the denuded mountainside; the voice of the soil in the crunch of alkali beneath the boots of a farmer pacing his land ruined by bad irrigation; the voice of the water as it roars crystal clear through the penstocks of a mighty dam, leaving behind the nutrients that once nourished a great floodplain and fed a vast fishery; the voice of the wind as it hisses with a load of dust whipped from the topsoil of half a county. Yes, the voices of Nature are quite easy to hear—if we will only listen. The question is, Will we? And if we do, Can we overcome our corrupt ways and marshal our efforts to collaborate with Nature as her brightest child and shepherd of Earth's life system? If not, it is almost certain that things will worsen for Nature, but even more so for us. Then, at a certain point in time we may no longer be able to cope with the adversity and we will perish. But Nature will survive, and so, too, will the insects, her most successful children. And as a final bit of irony, it will be insects that polish the bones of the very last of us to fall.

GLOSSARY

agro-ecosystem: The ecosystem composed of cultivated land and its adjacent or intermixed uncultivated surroundings, the plants contained or grown thereon, and the animals associated with those plants.

biological control: The regulation of plant and animal populations by natural enemies. The term is also applied to the practice of using natural enemies to control pests.

biocide: A chemical pesticide that is toxic to a wide range of species; e.g., insects, snails, birds, people.

biosphere: All the living organisms on Earth and their interacting, non-living environments.

bug: Technically, any insect of the order Hemiptera, but in popular language an insect of almost any kind, and for that matter other creepy-crawly cryptic things, too.

carbamate insecticide: An ester of carbamic acid having insecticidal properties (anti-choline-esterase activity).

cultural control: The application of agronomic, agricultural, and silvicultural practices, etc., to control pests; e.g., plowing-under of crop residues, pruning and destruction of infected tree branches, crop rotation.

economic threshold (= action threshold or action level). The density at which a pest population causes sufficient losses to justify the cost of control efforts.

frass: The refuse or excrement left by insect larvae.

hormone: A substance produced by the cells in one part of an organism's body and transported to another part of the body, where it produces a specific effect.

host: The organism that serves as the food source for a parasite or predator.

integrated control: A strategy that utilizes knowledge, monitoring, action criteria, materials, and methods, in concert with natural mortality factors, to manage pest populations.

integrated pest management: The same thing as integrated control.

land-grant university: An institution of higher education receiving federal aid under the Morrill Acts of 1862 and 1890. The name derives from the fact that under the original law each state was granted public lands for the support of at least one college teaching branches of learning related to agriculture and mechanic arts. As currently constituted, the typical land-grant university embraces a college of agriculture which includes an agricultural experiment station and a cooperative extension unit, and it conducts teaching as well.

natural control: The collective actions, of physical and biotic factors in the environment, that maintain species populations within characteristic limits.

natural enemy: An organism that causes the premature death of another organism.

organochlorine insecticide: A hydrocarbon with a certain number of chlorines which give it insecticidal properties (nerve-poison activity).

organophosphate insecticide: An ester of phosphoric acid with insecticidal properties (mostly anti-choline-esterase activity).

parasite: A small organism that lives and feeds in or on a larger, host organism.

pathogen: A micro-organism (microbe) that lives and feeds (parasitically) on or in a larger, host organism and thereby injures it.

pest: A species that, because of its great numbers, behavior, or feeding habits, is able to inflict substantial harm on man or his valued resources.

pest management: The manipulation of pest or potential-pest populations so as to diminish their injury or render them harmless. Pest management may involve simple manipulations such as spraying a rosebush or the emptying of water-filled tin cans to prevent mosquito breeding, or it may be effected through a complicated integrated control system.

pesticide resistance: Genetically selected tolerance of pest populations to pesticides, brought about by the pests' repeated exposure to chemical treatments.

pheromone: A substance secreted to the outside (of its body) by an individual of a species, which when received by another individual of its own species excites a specific response; e.g., mating response, aggregation response.

predator: An animal that feeds upon other animals (prey) that are either smaller or weaker than itself. The term is also sometimes applied to plant-feeding animals.

secondary pest outbreak: The eruption of a previously harmless species to injurious abundance in the wake of pesticide use that eliminates its natural enemies.

spider mite: A tiny, plant feeding, spider-like organism.

synthetic organic insecticide: A laboratory-synthesized carbon compound with insecticidal activity.

target-pest resurgence: The rapid resurgence of a chemically treated pest population to damaging abundance, brought about by the destruction, by the chemical treatment, of the natural enemies that otherwise would restrain the pest.

NOTES AND REFERENCES

1. N. E. Borlaug (undated). "Mankind and civilization at another crossroad," 1971 McDougall Memorial Lecture, presented on November 8, 1971, to the Seventh Biennial Conference of the Food and Agriculture Organization of the United Nations, Rome, Italy. Reproduced by American Breeders Service, Allis-Chalmers Corporation, J. I. Case Company, and Oscar Mayer & Co. Distributed by Wisconsin Agri-Business Council Inc. 48 pp.
2. E. F. Riek, 1970. "Fossil history," Ch. 8 of *The Insects of Australia.* Commonwealth Scientific and Industrial Research Organization (CSIRO), Canberra, Australia. Melbourne University Press. 1029 pp.
3. CSIRO, 1970. Page 1 of *The Insects of Australia.* Commonwealth Scientific and Industrial Research Organization (CSIRO), Canberra, Australia. Melbourne University Press. 1029 pp.
4. L. H. Newman, 1965. *Man and Insects,* p. 136. The Natural History Press. 252 pp.
5. Based on calculations made by my colleagues Kenneth S. Hagen and Richard Tassan, two of the world's most knowledgeable students of lady beetles and their feeding habits.
6. G. P. Georghiou, 1972. "The evolution of resistance to pesticides," *Annual Review of Ecology and Systematics* 3:133–68.
7. P. L. Adkisson, 1971. "Objective use of insecticides in agriculture," pp. 43–51 of *Agricultural Chemicals—Harmony or Discord for Food, People, Environment,* J. E. Swift (ed.). Univ. of California Div. of Agric. Sciences. 151 pp.
8. Instituto Centroamericano de Investigación y Tecnología Industrial (ICAITI), 1976. An environmental and economic study of the con-

sequences of pesticide use in Central American cotton production (final report of Phase I). Guatemala City, Guatemala, C.A., January 1976. 222 pp.

A. E. Olszyna-Marzys, M. de Campos, M. Taghi Farvar, and M. Thomas, 1973. "Chlorinated pesticide residues in human milk in Guatemala" (Spanish w/English summary). *Boletín de la Oficina Sanitaria Panamericana* 74(2): 93–107.

L. A. Falcon and R. Daxl, 1973. "Report to the government of Nicaragua on the integrated control of cotton pests for the period June 1970 to June 1973." UNDP. FAO (NIC/20/002/AGP) 1973. 60 pp.

A. Sequeira, R. Daxl, M. Sommeijer, A. van Huis, F. Pederson, O. León, and P. Giles, 1976. *Guía de Control Integrado de Plagas de Maís, Sorgo, y Frijol.* Min. Agric. y Ganad. Org. Nae. Unidas para la Agric. y Aliment. (FAO). Managua, Nicaragua, C.A. March 1976. 47 pp.

9. R. M. Hawthorne, 1970. *Estimated damage and crop loss caused by insect/mite pests.* California Dept. of Agriculture. E-82-13. Nov. 6, 1972. 12 pp.

10. R. F. Luck, R. van den Bosch, and R. Garcia, 1977. "*Chemical insect control, a troubled pest-management strategy.*" BioScience. 27:606–11

11. Anonymous. *Time,* July 12, 1976, pp. 38–46. "The bugs are coming."

12. R. M. Hawthorne, 1975. *Estimated damage and crop losses caused by insect-mite pests 1974.* State of California Dept. of Food and Agriculture. E-82-14. Sept. 4, 1975. 14 pp.

13. D. Pimentel, E. C. Terhune, W. Dritschilo, D. Gallanhan, N. Kinner, D. Nafus, R. Peterson, N. Zareh, J. Misiti, and O. Haber-Scham, 1977. "Pesticides, insects in food, and cosmetic standards," *BioScience* 27:178–85.

14. T. H. Jukes, 1972. "Scientific agriculture at the crossroads," *Agrichemical Age,* Nov. 1972, pp. 9–11.

Anonymous. "Who's who in the pesticide movement," *Agrichemical Age,* September–October 1975, pp. 9, 11.

I also have other documents in hand relating to the matter, which I prefer not to cite for fear that such citation might expose individuals to reprisal.

15. For example, the Velsicol Corporation and six persons are currently under indictment by a federal grand jury on charges they conspired to conceal potential cancer-causing effects of the insecti-

cides heptachlor and chlordane. I am also aware of an unpublicized case in which a chemical company attempted to conceal information on the accumulation of one of its experimental insecticides in fish taken from a treated lake. A recent article by Joann S. Lublin in *The Wall Street Journal* (February 21, 1978, pp. 1, 5) casts further doubt on the adequacy of pesticide safety tests.

16. A letter, with attachment, from E. Buyckx, entomologist, Plant Protection Service, FAO, Rome, to Ray F. Smith, dated February 9, 1972.
K. P. Shea, 1974. "Nerve damage," *Environment* 16 (9):6–9.
I also discussed this incident with several Egyptian entomologists, who stated that there were some human deaths in the episode.
17. P. Milius and D. Morgan, 1976. "How a toxic pesticide was kept off the market," Washington *Post*, Sunday, December 26, 1976.
18. *Forestry Chronicle* 51(4), August 1974. The entire issue was devoted to the spruce budworm problem.
19. See reference to ICAITI under Item 8.
20. U. S. Department of State. Telegram, unclassified, 2632, October 1, 1974. "Pesticide episode in Coahuila, Mexico . . . 689 cases of phosphate poisoning were reported, of which seven victims died."
21. State of California Department of Health and Department of Food and Agriculture, 1975. "Illnesses of employed persons reported by physicians to the state as due to exposure to pesticides, or their residues, according to type of illness." 1 page. (Kindly provided by Dr. Ephraim Kahn, chief, Epidemiological Studies Laboratory, California Dept. of Health.)
22. Dr. Kahn made this comment in a seminar held in the Biological Sciences Department, Stanford University, on May 3, 1976. During his discussion he presented data supportive of his conclusion.
23. Anonymous, 1970. Special Joint Meeting of the Tulare, Kings, and Delta Mosquito Abatement districts, 1737 West Huston Ave., Visalia, Calif., Sept. 23, 1970. 13 pp.
24. E. L. Baker, W. McWilson, M. Zeck, R. D. Dobbin, J. W. Miles, and S. Miller, 1977. *Malathion intoxication in spray workers in The Pakistan Malaria Control Program.* A report to the U. S. Agency for International Development, Department of State. U. S. Dept. of Health, Education & Welfare, Public Health Service Center for Disease Control, Atlanta, Ga. 13 pp. plus figures and tables.
25. "Malaria on the March," *Time*, Dec. 1, 1975, pp. 63–64.
M. A. Farid, 1975. "The world malaria situation." An address presented at the UNEP/WHO meeting on the Bio-Environmental

Methods of Control of Malaria, Lima, Peru, December 10–15, 1975. (Dr. Farid kindly granted me permission to cite his speech.)
G. P. Georghiou, 1972. "Studies on resistance to carbamate and organophosphorous insecticides in *Anopheles albimanus*," *Amer. Jour. Tropical Med. and Hygiene* 21:797–906.

26. A. I. Bischoff, 1969. *Azodrin wildlife investigations in California* (1969). Calif. Dept. of Fish and Game. Pesticides Project FW-1-R (mimeo).
27. M. Burman and E. DeHart, Jan. 31, 1972. Miami *Herald.*
28. Oakland (Calif.) *Tribune,* April 5, 1974.
Berkeley (Calif.) *Daily Gazette,* April 9, 1974.
29. R. L. Mull, 1972. *Excessive Guthion® residues in alfalfa hay.* Cooperative Extension, Univ. of California, June 1972. 1½ pp.
30. Memo from M. Hugh Handly, Kings Cty., Calif., agricultural commissioner and sealer, to all pest-control operators, pesticide dealers, and growers, dated June 3, 1975: "15 to 20 per cent of the samples of alfalfa hay treated with Supracide this season [1975] have been found to contain illegal residue." Quoted from a statement by John Hillis of the California Dept. of Food and Agriculture.
31. K. Burke, 1974. Los Angeles *Times,* August 13, 1974, Pt. II, p. 8.
C. M. Fennell, (agricultural commissioner, County of Imperial, Calif.). News release, April 14, 1975.
32. California Dept. of Food and Agriculture, 1974. *Apiary Inspectors' Newsletter.* Summary 1974. 2 pp.
Anonymous, 1975. "Apiary inspector says bee colony quality is up," *Western Hay and Grain Grower,* Oct. 1975, pp. 16, 18.
33. M. Smotherman. Personal communication.
34. R. van den Bosch, 1971. "Insecticides and the law," *The Hastings Law Journal* 22:615–28.
35. J. A. McMurtry, C. B. Huffaker, and M. van de Vrie, 1970. "Ecology of tetranychid mites and their natural enemies—a review. I. Tetranychid enemies: the biological characters and the impact of spray practices," *Hilgardia* 40:331–90.
C. B. Huffaker, M. van de Vrie, and J. A. McMurtry, 1970. "Ecology of tetranychid mites and their natural enemies—a review. II. Tetranychid populations and their possible control by predators: an evaluation," *Hilgardia* 40:391–458.
36. See R. M. Hawthorne, 1975 (Item 12).
37. C. Murray, 1976. *Chemical and Engineering News,* Feb. 2, 1976, pp. 17–18.

38. *Conservation News*, 1975. "Louisiana pelican deaths: more pesticides than data." Volume 40(16): 5–7.
39. R. F. Smith and H. T. Reynolds, 1972. "Effects of manipulation of cotton agro-ecosystems on insect pest populations," Chapter 21, pp. 373–406, of *The Careless Technology*, M. T. Farvar and J. P. Milton (eds.). Natural History Press. 1030 pp.
40. T. Boza Barducci, 1972. "Ecological consequences of pesticides used for the control of cotton insects in Cañete Valley, Peru," Ch. 23, pp. 423–38, of *The Careless Technology*, M. T. Farvar and J. P. Milton (eds.). Natural History Press. 1030 pp.
41. See ICAITI report under Item 8. Also see E. Eckholm 1977. "Unhealthy jobs," *Environment* 19(6): 29–38.
42. M. Taghi Farvar. Personal communication.
The high DDT levels are also discussed in A. E. Olszyna-Marzys et al., under Item 8.
43. See ICAITI report, under Item 8.
44. See ICAITI report and Sequeira et al., under Item 8.
45. L. A. Falcon and R. Daxl, 1973. "Report to the government of Nicaragua on the integrated control of cotton pests" (NIC/70/002/AGP) for the period June 1970 to June 1973. United Nations Development Programme, Food and Agriculture Organization, 1973. 60 pp.
46. See Item 7.
47. See Item 7.
48. At the inception of this program a letter cosigned by Ray F. Smith, Harold T. Reynolds, John E. Swift, and Robert van den Bosch was sent to the University of California dean (state-wide) of agricultural sciences, warning of the probable failure of the eradication program and stating that in our opinion the money and effort to be expended could be better used for research. We asked that this message be transmitted to the California Department of Agriculture, but I do not know whether this was done. Whatever the case, the letter seemingly had no effect, as the ill-fated eradication effort was undertaken despite our warning.
49. H. T. Reynolds, 1971. "A world review of the problem of insect population upsets and resurgences caused by pesticide chemicals," pp. 108–12 of *Agricultural Chemicals–Harmony or Discord for Food, People, Environment*. J. E. Swift (ed.). Univ. of California Div. of Agric. Sciences. 151 pp.
50. This information was gained in discussions of the cotton pest-control problem in Australia with a number of entomologists including

Americans R. F. Smith, T. F. Leigh, and H. T. Reynolds, and Australians D. E. Waterhouse, J. L. Readshaw, and N. Thompson.

51. This remark was made to me by a colleague in reaction to public statements I had made on the pesticide issue. Another colleague told his class, "Van den Bosch tells half-truths," while he and the class were discussing an article I had written for the *Daily Californian.*
52. This remark was blurted out by a colleague in a heated argument.
53. This remark was contained in a letter from a farm adviser who was embarrassed because I had published an article in *Organic Gardening and Farming* magazine.
54. Brawley *News*, Brawley, Calif., March 20, 1975, p. 1.
55. From a letter written by another farm adviser upset by one of my published articles.
56. D. Beeler, 1973. *Agrichemical Age*, January 1973, p. 22.
57. G. L. Berg, 1970. *Farm Chemicals*, September 1970.
58. This statement was made by an agricultural commissioner to several of my associates who were interviewing him regarding produce standards.
59. The Azodrin® production facility was damaged by fire between the 1965 and 1966 growing seasons, which severely limited output during 1966. However, the plant was restored to full capacity in time to meet the needs of the 1967 market season.
60. R. van den Bosch, 1969. "The toxicity problem—comments of an applied insect ecologist." Chapter 6 of *Chemical Fallout*, M. W. Miller and G. G. Berg (eds.). Charles C. Thomas Publ. 531 pp.
61. See Item 1.
62. I have had several conversations with Dr. Newsom concerning his recurrent clashes with the chemical companies. The account that I have given is essentially a distillate of these conversations. Dale Newsom is a remarkable person: an outstanding scientist, a man of complete honesty who will battle for principle, and a person of warmth and good humor. We have often differed in matters of approach and style, but we are in strong agreement on a number of points regarding the problems of contemporary pest control. Life would be easier if there were more Dale Newsoms in the pest-control field.
63. Letter from Denzel Ferguson to Robert van den Bosch, dated March 30, 1973.
64. F. Graham, Jr., 1970. *Since Silent Spring*. Fawcett Publ. pp. 158–59.

Dr. Rudd has also related this story of harassment to me, as have certain of his closest associates.

65. Letter from Charles Lincoln to Robert van den Bosch, dated April 2, 1973.
66. This incident occurred during the joint meeting of the Entomological Society of America and the Society's Southeastern Branch, in Miami, Florida, Nov. 30 to Dec. 3, 1970. Several persons directly involved in this incident related the details to me and to others.
67. This information was related to me by Theo F. Watson and Leon More, entomologists with the University of Arizona.
68. An undated, handwritten note to Dr. R. van den Bosch from Bob Fleet. This note was accompanied by a typewritten account of Fleet's tribulations.
69. N. Wade, 1973. "Agriculture: NAS panel charges inept management, poor research," *Science* 179:45–47.
A lot of static developed out of this report, and committee chairman Glenn S. Pound was severely criticized by certain of his peers. Interestingly, a subsequent report compiled by a different investigating team came up with essentially the same conclusions. This latter report was discussed in *Science*, Vol. 190, p. 959, in an article entitled "Agriculture: Academy group suggests major shakeup to President Ford."
70. National Research Council, National Academy of Sciences–National Academy of Engineering, 1967. *Report of Committee on the Imported Fire Ant, to Administrator, Agricultural Research Service, U. S. Department of Agriculture.* Contract No. 12-14-100-9447 (81). Sept. 28, 1967. 15 pp. Members of the panel were Harlow B. Mills, University of Wisconsin (Racine), chairman; F. S. Arant, Auburn University; M. F. Baker, U. S. Forest Service, Provo, Utah; W. S. Creighton, College of the City of New York (retired); J. C. Headley, University of Missouri; A. C. Hodson, University of Minnesota; R. C. Jung, director of health, New Orleans; J. W. Leonard, University of Michigan; R. March, University of California, Riverside; L. D. Newsom, Louisiana State University; Ray F. Smith, University of California, Berkeley; and B. V. Travis, Cornell University.
71. I learned the cost of this program at a meeting of the Wine Institute in San Francisco on Jan. 24, 1975, when a delegation of entomologists from the California Department of Food and Agriculture asked the Institute to apply pressure on the state legislature to restore state general-fund support for the $660,000 program.

72. Anonymous, 1965. *Bull. Ent. Soc. America* 11(1):33.
73. F. E. Egler, 1964. "Pesticides in our ecosystem. Communication II," *BioScience* 14(11): 29–36.
74. L. C. Cole, 1965. "Pesticides, petulance, post-mortem and pax" (a letter to *BioScience*), *BioScience* 15(2): 158–59.
75. Entomological Society of America president (for 1976) Ray F. Smith told me that the Society's Sustaining Associates (essentially a group of chemical companies) had originally promised to deliver Norman Borlaug as distinguished speaker but failed to land their prize and so produced Long as a substitute.
76. Letter from Sen. Walter F. Mondale to Robert van den Bosch, dated March 12, 1969.
77. See Item 1.
78. B. E. Wickman, R. R. Mason, and C. G. Thompson, 1973. *Major outbreaks of the Douglas fir tussock moth in Oregon and California.* USDA Forest Service Technical Report PNW-5. 18 pp.
79. R. Davis, 1974. "The politics of DDT," *Defenders of Wildlife* 49:582–84.
80. F. Graham, 1975. "Update: Moths and DDT." *Audubon* 77(2):120–21.
81. S. G. Herman, 1974. "DDT fiasco," *Defenders of Wildlife* 49:449–54.
82. *Western Livestock Journal, Western Edition*, Vol. 53(6):1, Dec. 2, 1974.
83. Seattle *Post-Intelligencer*, Nov. 23, 1974.
84. See Item 78.
85. Personal communication from Dr. Carroll B. Williams, Jr., Pacific Southwest.
86. Cited in Item 77.
87. State of Louisiana. *A Petition* [to EPA] *for Emergency Use of DDT on Cotton Against the Tobacco Budworm.* Jan. 14, 1974.
88. C. B. Williams, Jr., 1974. Seminar presented before the Department of Entomology, Univ. of California, Davis, Nov. 27, 1974.
89. Every Forest Service research entomologist and land-grant-university forest entomologist with whom I discussed this matter made essentially the same assessment of the problem.
90. S. G. Herman, 1973. "The present situation," pp. 7–16 of *Douglas-fir Tussock Moth in the Northwest: The Case Against Use of DDT in 1974. Ecology and Chemistry of Pollution.* The Evergreen State College, Olympia, Wash. 25 pp.
91. This information was provided to the Finance Committee of the California State Senate in September 1977 by the California Department of Food and Agriculture.

92. M. Brown, R. Garcia, C. Magowan, A. Miller, M. Moran, D. Pelzer, J. Swartz, and R. van den Bosch, 1977. *Investigation of the effects of food standards on pesticide use.* A report to the U. S. Environmental Protection Agency, prepared under EPA Contract 68-01-26021. 160 pp.

93. In this book I have several times criticized the University of California, especially as regards the affinity or subservience of a number of its high administrators to the power elite of agri-business and other vested interests. My gripes are largely those of a shy lover, for, in actuality, despite her warts and wrinkles, the University of California is in my estimation the finest public institution of higher learning that our society has created. The greatness of the University lies in its unique charter, which permits it to function as a virtually autonomous corporation under the stewardship of its Board of Regents, who have had the wisdom to leave most academic affairs to the Academic Senate.

In this favorable climate, I have been able to play the role of eco-radical without constraint. And in this connection I am compelled to voice a word of respect for University Vice-President for Agricultural Sciences James B. Kendrick, Jr. (an unabashed proponent of agri-business), who, I know, is often embarrassed and exasperated by my antics. Kendrick has several times told me that he receives frequent complaints about me from powerful agri-business interests. Others have let me know that some of the complainers have exhorted Kendrick to throttle me, but although he deplores my antics, Kendrick has assured me that as long as I deal in the truth he will not interfere with my activities.

Jim Kendrick and I, though cats with different stripes, both understand and respect the University of California, and thus we survive.

94. J. Walsh, 1976. "Cosmetic standards: are pesticides overused for appearance's sake?" *Science* 193:744–47.

95. This information is contained in a brochure prepared by Pest Management Associates, 250 Lincoln St., West Lafayette, Ind. 47906. Officers listed are F. T. Turpin, president; A. P. Gutierrez, vice-president; W. R. Campbell, secretary; R. T. Huber, treasurer; and R. C. Hall, assistant treasurer.

I have also discussed the corn rootworm problem in considerable detail with Drs. Turpin and Gutierrez.

96. United States Environmental Protection Agency, Office of Pesticide Programs—Office of Water and Hazardous Materials, 1974. *Farmers' pesticide use decisions and attitudes on alternate crop protection methods.* EPA-540/1-24-002. July 1974. 157 pp. and Appendices.

97. Included as an enclosure in a letter sent to Dr. John E. Swift on September 3, 1974, by Charles A. Black, executive vice-president of CAST.
98. J. M. Witt, 1976. The contrary statement to–Proposed that the ESA (Entomological Society of America) should affiliate with CAST. *Bull. Ent. Soc. America* 22(1): 31–36.
99. University Bulletin (Univ. of California) Vol. 21: 154, May 28, 1973: "News from the Campuses. Awards and Honors. Jukes, Thomas H. Professor of Medical Physics in Residence, Berkeley: Honorary Recognition for Communicative Skills from the Chevron Chemical Company."
100. See Item 9.
101. See Item 96.
102. *Agrichemical Age*, Jan.–Feb. 1976, p. 46.
103. *Agrichemical Age*, Jan.–Feb. 1976, p. 4.
104. See Item 34.
105. The presentation was made on December 1, 1975, under the title "A report to ESA on the status of entomology in the People's Republic of China." The discussion leader was Dr. Robert L. Metcalf, whose observations form the basis of most of my remarks. Recently a synopsis of Metcalf's speech appeared in *Environment*, 18(9):14–17. A full report of the visit has now been published by the National Academy of Sciences as CSCPRC Report No. 2 (1977), *Insect Control in the People's Republic of China*. 218 pp.
106. A. D. Telford, 1971. *Annual Report, Marin County* [Calif.] *Mosquito Abatement District #1. 1971*, p. 6.
I have also discussed details of the program with Dr. Telford.
107. W. Olkowski, H. Olkowski, R. van den Bosch, and R. Hom, 1976. "Ecosystem management: A framework for urban pest control," *BioScience* 26:384–89.
108. S. C. Hoyt and L. E. Caltagirone. 1971. "The developing program of integrated control of pests of apples in Washington and peaches in California," pp. 395–421 of *Biological Control*, C. B. Huffaker (ed.). Plenum Press. 511 pp.
109. D. C. Hall, R. B. Norgaard, and P. K. True, 1975. "The performance of independent pest-management consultants in San Joaquin cotton and citrus," *California Agriculture* 29(10):12–14.
110. These figures have either appeared in published reports or have been communicated to me by the concerned researchers.
111. See Item 13.

112. M. Harris, 1973. "The withering green revolution," *Natural History* 82(3):20–22.
113. G. L. Bush, R. W. Neek, and G. B. Kitto, 1976. "Screwworm eradication: inadvertent selection for noncompetitive ecotypes during mass rearing," *Science* 193:491–93.
114. Personal communication from Dr. W. G. Eden, who was a member of a two-man panel that investigated the feasibility of the expanded eradication program.

SELECTED BIBLIOGRAPHY

Carson, R. 1962. *Silent Spring*. Houghton Mifflin Co., Boston. 368 pp.

Council of Environmental Quality. 1972. *Integrated Pest Management*. U. S. Government Printing Office, Washington, D.C. 0-464-590. 41 pp.

DeBach, P. 1975. *Biological Control by Natural Enemies*. Cambridge University Press, London. 323 pp.

—— (ed.). 1964. *Biological Control of Insect Pests and Weeds*. Chapman & Hall, London. 844 pp.

Dethier, V. G. 1975. *Man's Plague? Insects and Agriculture*. Darwin Press, Princeton, N.J. 237 pp.

Farvar, M. T.; and Milton, J. P. (eds.) 1972. *The Careless Technology*. Natural History Press, Garden City, N.Y. 1030 pp.

Flint, M. L.; and van den Bosch, R. 1977. *A Source Book on Integrated Pest Management*. Department of Health, Education, & Welfare, Office of Environmental Education, Washington, D.C. 240 pp. (illustrated). [Available through the author (RvdB.) while they last.]

Glass, E. H. (co-ordinator). 1976. *Integrated Pest Management: Rationale, Potential, Needs, and Implementation*. Entomol. Soc. Amer. Special Publ., 75-2. 141 pp.

Graham, F., Jr. 1970. *Since Silent Spring*. Houghton Mifflin Co., Boston. 333 pp.

Harmer, R. M. 1971. *Unfit for Human Consumption*. Prentice-Hall, Englewood Cliffs, N.J. 374 pp.

Huffaker, C. B. (ed.). 1971. *Biological Control*. Plenum Press, New York. 511 pp.

——; and Messenger, P. S. (eds.). 1977. *The Theory and Practice of Biological Control.* Academic Press, New York. 788 pp.

Mellanby, K. 1967. *Pesticides and Pollution.* William Collins Sons, London. 219 pp.

Metcalf, R. L.; and Luckman, W. 1975. *Introduction to Insect Pest Management.* John Wiley & Sons, New York. 587 pp.

Rabb, R. L.; and Guthrie, F. E. 1970. *Concepts of Pest Management.* North Carolina State University, Raleigh. 242 pp.

Rudd, R. L. 1964. *Pesticides and the Living Landscape.* University of Wisconsin Press, Madison. 320 pp.

Smith, R. F.; and van den Bosch, R. 1967. "Integrated Control," Ch. 9, pp. 295–340, of *Pest Control–Biological, Physical, and Selected Chemical Methods.* W. W. Kilgore and R. L. Doutt (eds.). Academic Press, New York. 477 pp.

ACKNOWLEDGMENTS

Many persons contributed to the evolution of this little volume, and I regret that I cannot recall each and every name and contribution. However, those persons know me, and I feel certain that they understand that any oversight is unintentional.

But there are a number of helpers, advisers, and encouragers whom I do remember, and I thank them warmly. First, there is my wife, Peggy, who typed the initial draft and put up with me during its gestation. Then there is Christine Merritt, who helped immensely in documenting the book and also critically reviewed the manuscript. Nettie Mackey, Joanne Fox, and Patricia Felch, who typed the final manuscript, also have my lasting gratitude.

I am grateful, too, to Mary Louise Flint, Andrew P. Gutierrez, Robert Wuliger, Lloyd A. Andres, Shirley Briggs, Ray F. Smith, Carl B. Huffaker, and Paul Ehrlich, for constructively reviewing all or parts of the manuscript and for offering encouragement.

Finally, I thank Jane Clarkin for her many kindnesses.